D1474942

Cancer Cytogenetics

METHODS IN MOLECULAR BIOLOGY™

John M. Walker, Series Editor

234. **p53 Protocols**, edited by *Sumitra Deb and Swati Palit Deb, 2003*
233. **Protein Kinase C Protocols**, edited by *Alexandra C. Newton, 2003*
232. **Protein Misfolding and Disease:** *Principles and Protocols*, edited by *Peter Bross and Niels Gregersen, 2003*
231. **Directed Evolution Library Creation**: *Methods and Protocols*, edited by *Frances H. Arnold and George Georgiou, 2003*
230. **Directed Enzyme Evolution**: *Screening and Selection Methods*, edited by *Frances H. Arnold and George Georgiou, 2003*
229. **Lentivirus Gene Engineering Protocols**, edited by *Maurizio Federico, 2003*
228. **Membrane Protein Protocols**: *Expression, Purification, and Characterization*, edited by *Barry S. Selinsky, 2003*
227. **Membrane Transporters:** *Methods and Protocols,* edited by *Qing Yan, 2003*
226. **PCR Protocols, Second Edition,** edited by *John M. S. Bartlett and David Stirling, 2003*
225. **Inflammation Protocols,** edited by *Paul G. Winyard and Derek A. Willoughby, 2003*
224. **Functional Genomics:** *Methods and Protocols,* edited by *Michael J. Brownstein and Arkady Khodursky, 2003*
223. **Tumor Suppressor Genes:** *Volume 2: Regulation, Function, and Medicinal Applications,* edited by *Wafik S. El-Deiry, 2003*
222. **Tumor Suppressor Genes:** *Volume 1: Pathways and Isolation Strategies,* edited by *Wafik S. El-Deiry, 2003*
221. **Generation of cDNA Libraries:** *Methods and Protocols,* edited by *Shao-Yao Ying, 2003*
220. **Cancer Cytogenetics:** *Methods and Protocols,* edited by *John Swansbury, 2003*
219. **Cardiac Cell and Gene Transfer:** *Principles, Protocols, and Applications,* edited by *Joseph M. Metzger, 2003*
218. **Cancer Cell Signaling:** *Methods and Protocols,* edited by *David M. Terrian, 2003*
217. **Neurogenetics:** *Methods and Protocols,* edited by *Nicholas T. Potter, 2003*
216. **PCR Detection of Microbial Pathogens:** *Methods and Protocols,* edited by *Konrad Sachse and Joachim Frey, 2003*
215. **Cytokines and Colony Stimulating Factors:** *Methods and Protocols,* edited by *Dieter Körholz and Wieland Kiess, 2003*
214. **Superantigen Protocols,** edited by *Teresa Krakauer, 2003*
213. **Capillary Electrophoresis of Carbohydrates,** edited by *Pierre Thibault and Susumu Honda, 2003*
212. **Single Nucleotide Polymorphisms:** *Methods and Protocols,* edited by *Pui-Yan Kwok, 2003*
211. **Protein Sequencing Protocols, Second Edition,** edited by *Bryan John Smith, 2003*
210. **MHC Protocols,** edited by *Stephen H. Powis and Robert W. Vaughan, 2003*
209. **Transgenic Mouse Methods and Protocols,** edited by *Marten Hofker and Jan van Deursen, 2003*
208. **Peptide Nucleic Acids:** *Methods and Protocols,* edited by *Peter E. Nielsen, 2002*
207. **Recombinant Antibodies for Cancer Therapy:** *Methods and Protocols,* edited by *Martin Welschof and Jürgen Krauss, 2002*
206. **Endothelin Protocols,** edited by *Janet J. Maguire and Anthony P. Davenport, 2002*
205. ***E. coli*** **Gene Expression Protocols,** edited by *Peter E. Vaillancourt, 2002*
204. **Molecular Cytogenetics:** *Protocols and Applications,* edited by *Yao-Shan Fan, 2002*
203. ***In Situ*** **Detection of DNA Damage:** *Methods and Protocols,* edited by *Vladimir V. Didenko, 2002*
202. **Thyroid Hormone Receptors:** *Methods and Protocols,* edited by *Aria Baniahmad, 2002*
201. **Combinatorial Library Methods and Protocols,** edited by *Lisa B. English, 2002*
200. **DNA Methylation Protocols,** edited by *Ken I. Mills and Bernie H. Ramsahoye, 2002*
199. **Liposome Methods and Protocols,** edited by *Subhash C. Basu and Manju Basu, 2002*
198. **Neural Stem Cells:** *Methods and Protocols,* edited by *Tanja Zigova, Juan R. Sanchez-Ramos, and Paul R. Sanberg, 2002*
197. **Mitochondrial DNA:** *Methods and Protocols,* edited by *William C. Copeland, 2002*
196. **Oxidants and Antioxidants:** *Ultrastructure and Molecular Biology Protocols,* edited by *Donald Armstrong, 2002*
195. **Quantitative Trait Loci:** *Methods and Protocols,* edited by *Nicola J. Camp and Angela Cox, 2002*
194. **Post-translational Modification Reactions,** edited by *Christoph Kannicht, 2002*
193. **RT-PCR Protocols,** edited by *Joseph O'Connell, 2002*
192. **PCR Cloning Protocols, Second Edition,** edited by *Bing-Yuan Chen and Harry W. Janes, 2002*
191. **Telomeres and Telomerase:** *Methods and Protocols,* edited by *John A. Double and Michael J. Thompson, 2002*
190. **High Throughput Screening:** *Methods and Protocols,* edited by *William P. Janzen, 2002*
189. **GTPase Protocols:** *The RAS Superfamily,* edited by *Edward J. Manser and Thomas Leung, 2002*
188. **Epithelial Cell Culture Protocols,** edited by *Clare Wise, 2002*
187. **PCR Mutation Detection Protocols,** edited by *Bimal D. M. Theophilus and Ralph Rapley, 2002*
186. **Oxidative Stress and Antioxidant Protocols,** edited by *Donald Armstrong, 2002*
185. **Embryonic Stem Cells:** *Methods and Protocols,* edited by *Kursad Turksen, 2002*
184. **Biostatistical Methods,** edited by *Stephen W. Looney, 2002*
183. **Green Fluorescent Protein:** *Applications and Protocols,* edited by *Barry W. Hicks, 2002*
182. **In Vitro Mutagenesis Protocols, 2nd ed.**, edited by *Jeff Braman, 2002*
181. **Genomic Imprinting:** *Methods and Protocols,* edited by *Andrew Ward, 2002*

METHODS IN MOLECULAR BIOLOGY™

Cancer Cytogenetics

Methods and Protocols

Edited by

John Swansbury

The Royal Marsden NHS Trust, The Institute of Cancer Research, Sutton, Surrey, UK

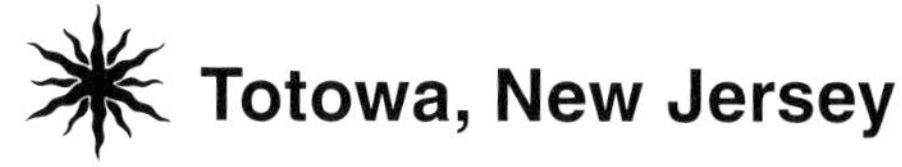

Humana Press Totowa, New Jersey

999 Riverview Drive, Suite 208
Totowa, New Jersey 07512

www.humanapress.com

This publication is printed on acid-free paper. ∞
ANSI Z39.48-1984 (American Standards Institute)
Permanence of Paper for Printed Library Materials.

Cover illustration: Fig. 1A from Chapter 11, "Human Solid Tumors," by Pietro Polito, Paola Dal Cin, Maria Debiec-Rychter, and Anne Hagemeijer.

Production Editor: Tracy Catanese
Cover design by Patricia F. Cleary.

For additional copies, pricing for bulk purchases, and/or information about other Humana titles, contact Humana at the above address or at any of the following numbers: Tel.: 973-256-1699; Fax: 973-256-8341; E-mail: humana@humanapr.com; or visit our Website: www.humanapress.com

Printed in the United States of America. 10 9 8 7 6 5 4 3 2 1

Library of Congress Cataloging in Publication Data

Cancer cytogenetics : methods and protocols / edited by John Swansbury.
p. ; cm. — (Methods in molecular biology ; 220)
Includes bibliographical references and index.
ISBN 1-58829-080-8 (alk. paper); 1-59259-363-1 (e-book)
1. Cancer--Genetic aspects--Laboratory manuals. I. Swansbury, John. II. Series.

RC268.4 .C349 2003
616.99'4042—dc21

2002027288

Preface

During the past 20 years, as genetic data have been shown to be intimately associated with diagnosis and prognosis, cytogenetic studies of malignancy have moved firmly out of research laboratories and into the front line of clinical practice. In almost every kind of hematologic malignancy, for example, it has been found that the type of clonal abnormality present at diagnosis is either the most important prognostic factor or one of the most important. The same is being found to apply in most solid tumors, and new prognostic associations are constantly being discovered. Advances in treatment have improved the chances of survival in many kinds of cancer, but have not diminished the differential prognostic effect of knowing what genetic abnormalities are present. Furthermore, the role of genetics and cytogenetics in medicine is likely to expand greatly as the findings of the Human Genome Project start to have effect.

It is no wonder, therefore, that there is a huge worldwide increase in the demand for cytogenetic studies in malignancy. However, there are not enough experienced laboratories to cope with the many requests to train new cytogeneticists in this challenging branch of medical science. It is the author's hope that *Cancer Cytogenetics: Methods and Protocols* will help. It offers practical advice, not only on the technical aspects of performing diagnostic cytogenetic studies, but also on the interpretation of the findings, the establishment of a diagnostic service, and the training of staff. In this respect, its emphasis is slightly different from that of may other books in the *Methods in Molecular Biology*™ series, which generally concentrate on techniques. In cytogenetics, the techniques are relatively easy to learn, but analysis and interpretation call for a great deal more experience. This book consequently includes comments and advice on a wide variety of issues associated with the provision of a diagnostic cytogenetics service.

In some countries, there are formal, postgraduate training schemes for cytogeneticists; in the United Kingdom, for example, these studies take 2 years to complete. This thorough approach is warmly recommended because it ensures that the cytogeneticist acquires a wide experience in all aspects of the work. However, in other countries such schemes are not yet available, and training is often dependent on a placement in an established laboratory. The editor's laboratory has been host to visitors from around the world, with various scientific and medical backgrounds, who have endeavored to learn the essentials of malignancy cytogenetics so that they can start up their own service. Their enthusiasm and efforts have been a major part in the inspiration for *Cancer Cytogenetics*. It has been slanted, therefore, toward those who plan to start up a diagnostic (or research) malignancy cytogenetics service, whether their laboratory has previously done only constitutional studies, or has done no previous cytogenetics work at all. However, this book should be of interest even for those who already have extensive experience: no one has yet developed a technique for malignancy cytogenetics that will give the consistently successful, good-quality result that can be obtained in, for example, constitutional studies of PHA-stimulated lymphocytes.

By working through *Cancer Cytogenetics*, trying the techniques, and intelligently adapting them as necessary to suit local conditions, it should be possible to start to get useful results. No book can by itself provide the experience that can be gained by visiting a well-established laboratory; however, more benefit will be gained from such a visit if it is used to enhance what has already been learned from this book. It can be an expensive business for both the host and for the visitor when the visitor arrives knowing nothing at all.

A few very elementary chapters have been included to help those who have no prior knowledge of chromosome analysis and terminology. Other chapters may be of more interest to those who have already acquired some cytogenetics expertise and are contemplating extending their repertoire to include such complementary techniques such as CGH, FISH, and M-FISH. For a book with declared bias toward the novice, it may seem ambitious to have included chapters on these techniques, which require expensive, computer-based equipment. However, the companies that produce this equipment are not slow to place their systems in developing markets, and sometimes fail to emphasize the necessity of first being able to produce a conventional karyotype. In such instances, the development of a comprehensive cytogenetics service appears to flow in reverse, with the high-tech sophistication of FISH being established before the laboratory has any expertise in low-tech, basic chromosome analysis. However, it is risking disaster to try to interpret FISH and molecular results without knowing the fundamentals of normal and abnormal human karyotypes.

Several overview chapters are included to provide some information about the most common clinical and cytogenetic aspects of the diseases being described. For a more detailed text, that by Heim and Mitelman has not yet been surpassed ***(1)***.

Detailed and original descriptions of techniques have been published in scientific journals, but some of these appeared many years ago and are now no longer readily available. It seemed to the publishers that the subject might benefit from a more detailed approach, with scope for mention of the subtle variations in technique that laboratories use for different hematologic disorders, while retaining the clarity of a simple approach. This volume attempts to meet this aim. Some recommended books describing techniques for all kinds of cytogenetics studies, including those in malignancy, are *Analyzing Chromosomes* **(2)** and the extensive compendium of cytogenetic information and techniques produced by the American Association of Genetic Technologists **(3)**. However, even this large volume devotes rather little space to solving the particular problems associated with specific malignancies such as multiple myeloma and chronic lymphocytic leukemia, which are among the most difficult hematologic disorders to study by conventional cytogenetics. Other books, including some in the Humana series, have included a chapter on cytogenetic methods for studying human malignancy, but most of these have inevitably been of a rather general nature. A new edition of *Human Cytogenetics: Malignancy and Acquired Abnormalities* **(4)** does provide more comprehensive practical information.

I am grateful to those who have contributed chapters to *Cancer Cytogenetics*, all of whom are very experienced; the names of most of them will be well known to malignancy cytogeneticists. The institutions in which they work continue to make major contributions to our knowledge of genetic abnormalities in malignancy. Inevitably, in a book of this type with contributions from several authors, there is some overlap and duplication, both in the techniques being described and in the explanatory comments. Any differences are not contradictory, but are complementary; they are not intended to cause concern or confusion, but rather to illustrate the validity of different approaches. There is no one technique that will work perfectly on all samples in all places; it is important to know what variations can be tested to improve results locally.

The publication of *Cancer Cytogenetics* has been deliberately delayed to include an assessment of the practical role of the newest technologies that are being added to the cytogeneticist's repertoire, now that these are coming into routine use. The competition between the developing companies to produce a reliable system for multicolor fluorescence in situ hybridization (FISH) has been intense; each company has added to the plethora of similar terms and procedures which all aim to produce a similar result. These include multiplex FISH, multifluor FISH, spectral karyotyping, and RX-FISH. Although undeniably useful, the very variety causes some confusion to those who simply want to provide a clinical service on a limited budget. The range covered in this book has been restricted. For those who wish to take this subject further, a more comprehensive list of types of FISH is contained in *Molecular Cytogenetics: Protocols and Applications*, a recent volume of the *Methods in Molecular Biology*™ series **(5)**.

Many of these FISH-based technologies still need dividing cells; therefore the ability to obtain suitable divisions is a major preoccupation in the cytogenetics world. Other technologies have also become available that do not depend on obtaining dividing tumor cells from the tissue being studied. They are sometimes grouped together as "molecular" genetics tests and include Southern (and other) blots and other techniques based on polymerase chain reaction (PCR). It has been decided not to include any chapters describing how to perform them. This is principally because there is already an excellent recent book in the *Methods in Molecular Medicine*™ series devoted to molecular assays ***(6)***. Instead, this book will concentrate on conventional cytogenetics and the tests that are most closely associated. There are now so many ways of analyzing the genetic abnormalities present in tumor material that it can be difficult for the inexperienced cytogenetics laboratory director to know which to use. There are very few laboratories that have such generous funding and so many staff that all possible technologies can be brought into routine use. Therefore, throughout *Cancer Cytogenetics* are comments about the main advantages and disadvantages of each type of technology, in the hope that they will help with these difficult decisions.

Methodology is a constant topic of conversation among cytogeneticists, and I am indebted to the many who, often unaware, have contributed to this book by sharing experience, theories, and advice. I am also grateful to the Royal Marsden Hospital NHS Trust, under whose auspices this book was written, to Professor Catovsky for his encouragement, to Applied Imaging International for subsidizing the cost of the color FISH pictures, and to my wife and family for their support and patience.

John Swansbury

References

1. Heim, S., and Mitelman, F. (1995) *Cancer Cytogenetics* (2nd edition). A. R. Liss, New York, NY.
2. Czepulkowski, B. (2001) *Analyzing Chromosomes*. BIOS, Oxford, UK and New York, NY.
3. Barch, M. J., Knutsen, T., and Spurbeck, J. L. (1997) *The AGT Cytogenetic Laboratory Manual* (3rd edition). Lippincott-Raven, New York, NY.
4. Rooney, D. E. (2001) *Human Cytogenetics: Malignancy and Acquired Abnormalities* (3rd edition). IRL, Oxford, UK and Washington, DC.
5. Fan, Y.-S., ed. (2002) *Molecular Cytogenetics: Protocols and Applications*, Methods in Molecular Biology, vol. 204, Humana, Totowa, NJ.
6. Boultwood, J., ed. (2001) *Molecular Analysis of Cancer*, Methods in Molecular Medicine, vol. 68, Humana, Totowa, NJ.

Contents

Contributors

PAOLA DAL CIN • *Center for Human Genetics, Belgium*
MARIA DEBIEC-RYCHTER • *Center for Human Genetics, Belgium*
ANNE HAGEMEIJER • *Center for Human Genetics, Belgium*
SUSAN MATHEW • *Cytogenetics Laboratory, Department of Pathology, St. Jude Children's Research Hospital, Memphis, TN*
TOON MIN • *Academic Haematology and Cytogenetics, The Royal Marsden NHS Trust, Sutton, Surrey, England*
PIETRO POLITO • *Center for Human Genetics, Belgium*
SUSANA C. RAIMONDI • *Cytogenetics Laboratory, Department of Pathology, St. Jude Children's Research Hospital, Memphis, TN*
JON C. STREFFORD • *ICRF Medical Oncology Unit, John Vane Science Center, Queen Mary and Westfield College, London, UK*
JOHN SWANSBURY • *Academic Haematology and Cytogenetics, The Royal Marsden NHS Trust, The Institute of Cancer Research, Surrey, UK*
ANN WATMORE • *North Trent Cytogenetics Service, Sheffield Children's Hospital Trust, Sheffield, UK*

1

Introduction

John Swansbury

1. The Clinical Value of Cytogenetic Studies in Malignancy

The vast majority of published cytogenetic studies of malignancy have been of leukemias and related hematologic disorders (*see* **Fig. 1**), even though these constitute only about 20% of all cancers. It follows that most of what is known about the clinical applications of cytogenetic studies has been derived from hematologic malignancies. More recently, however, there has been a huge expansion in knowledge of the recurrent abnormalities in some solid tumors, and it is clear that in these, just as in the leukemias, cytogenetic abnormalities can help to define the diagnosis and to indicate clear prognostic differences. Consequently, cytogenetic studies of some solid tumors are now also moving out of the research environment and into routine clinical service.

If all patients with a particular malignancy died, or all survived, then there would be little clinical value in doing cytogenetic studies; they would have remained in the realm of those researchers who are probing the origins of cancer. Even as recently as 20 yr ago, cytogenetic results were still regarded by many clinicians as being of peripheral interest. However, in all tumor types studied so far,

From: *Methods in Molecular Biology, vol. 220: Cancer Cytogenetics: Methods and Protocols*
Edited by: John Swansbury © Humana Press Inc., Totowa, NJ

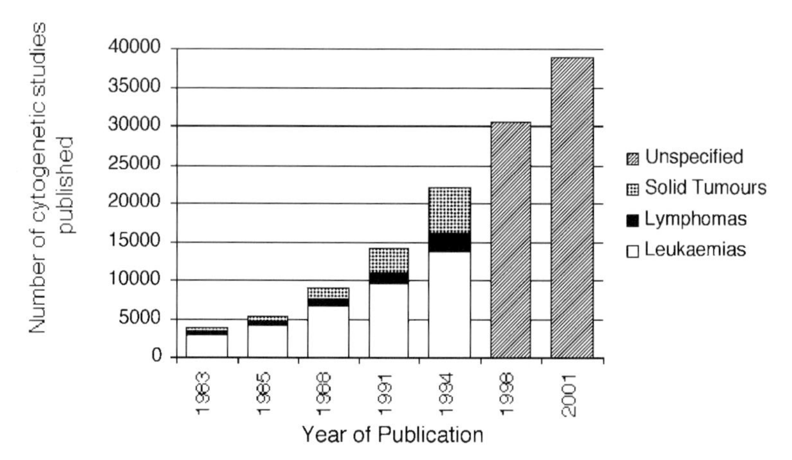

Fig. 1. Number of karyotypes published in successive *Mitelman's Catalogs of Chromosome Aberrations in Cancer*; data obtained directly from the catalogs. The 1998 edition was published on CD-ROM, and the current edition is online. Note that cases of chronic myeloid leukemia with a simple t(9;22)(q34;q11) were excluded, which therefore increases the overall number of published cases of leukemia.

the presence or absence of many of the genetic abnormalities found has been associated with different responses to treatment. Therefore, genetic and cytogenetic studies are being recognized as essential to the best choice of treatment for a patient. As a consequence of these advances, clinical colleagues now expect that cytogenetic analysis of malignancy will provide rapid, accurate, and specific results to help them in the choice of treatment and the management of patients. There is a greatly increased pressure on the cytogeneticist to provide results that fulfil these expectations. For example, at one time most patients with acute leukemia were given rather similar treatment for the first 28 d, and so there was little need to report a study in less than this time. Now treatment decisions for some patients with acute promyelocytic leukemia or Ph+ acute leukemia are made within 24 h. There is more to the management of a patient than merely choosing the most appropriate type of treatment, however: for every patient, and his or her family, the diagnosis of a malignancy can be traumatic, and an accurate and early indication of their prognosis is valuable.

2. Applications and Limitations of Conventional Cytogenetics Studies

It is helpful to be aware of the applications/strengths and the limitations/weaknesses of conventional cytogenetics, and to know when the use of other genetic assays may be more appropriate. A clinician may request a specific type of study, which may or may not be appropriate for the information sought. Conversely, the cytogeneticist may be asked to advise on the best approach. It is important for both parties to be aware of the likelihood of false-positive and false-negative results, and know what steps can be taken to minimize these.

2.1. Applications

The usual clinical applications of cytogenetic studies of acquired abnormalities in malignancy are:

1. To establish the presence of a malignant clone.
2. To clarify the diagnosis.
3. To indicate a prognosis.
4. To assist with the choice of a treatment strategy.
5. To monitor response to treatment.
6. To support further research.

These are considered in a little more detail in the following:

1. *To establish the presence of a malignant clone.* Detection of a karyotypically abnormal clone is almost always evidence for the presence of a malignancy, a rare exception being trisomies found in reactive lymphocytes around renal tumors (*see* Chapter 12). Demonstrating that there is a clone present is particularly helpful in distinguishing between reactive conditions and malignancy. Examples are investigating a pleural effusion, a lymphocytosis, or an anemia. However, it must always be remembered that the finding of only karyotypically normal cells does not prove that there is no malignant clone present. It may happen that all the cells analyzed came from normal tissue.
2. *To clarify the diagnosis.* Some genetic abnormalities are closely associated with specific kinds of disease, and this is particularly helpful

when the diagnosis itself is uncertain. For example, the small round-cell tumors, a group of tumors that usually occur in children, may be indistinguishable by light microscopy; other laboratory tests are needed to give an indication of the type of malignancy. Several of these tumors commonly have specific translocations, and these may be detected by fluorescene *in situ* hybridization (FISH) as well as by conventional cytogenetics (*see* Chapter 10).

A cytogenetic study can also help to distinguish between a relapse and the emergence of a secondary malignancy; this is described in more detail in Chapter 12. The type of cytogenetic abnormalities found can be significant: loss of a 5 or a 7 or partial deletion of the long arms of these chromosomes is most common 3 yr or more after previous exposure to akylating agents, and indicate a poor prognosis. Abnormalities of 11q23 or 21q22 tend to occur < 3 yr after exposure to treatment with topoisomerase II inhibitors, in which case the response to treatment is likely to be better. The finding of such abnormalities in a new clone that is unrelated to the clone found at first diagnosis is suggestive of a new, secondary malignancy rather than relapse of the primary.

Occasionally a child is born with leukemia; a cytogenetic study will help to distinguish between a transient abnormal myelopoiesis (TAM), which is a benign condition that will resolve spontaneously, most common in babies with Down syndrome, and a true neonatal leukemia, in which the most common cytogenetic findings are t(4;11)(q21;q23) or some other abnormality of 11q23, and which are associated with a very poor prognosis.

3. *To indicate prognosis, independently or by association with other risk factors.* In most kinds of hematologic malignancies, certain cytogenetic abnormalities are now known to be either the most powerful prognostic indicator, or one of the most important. This effect persists despite advances in treatment. The same effects are also being demonstrated in solid tumors. The presence of any clone does not automatically mean that the patient has a poor prognosis: some abnormalities are associated with a better prognosis than a "normal" karyotype and some with worse. Most cytogeneticists quite reasonably hesitate to use the word normal to describe a malignancy karyotype: because all cancer arises as a result of genetic abnormality, failure to find a clone implies either that the cells analyzed did not derive from the malignant cells, or that they did but the genetic abnormality was undetectable.

4. *To assist with the choice of a treatment strategy.* In many modern treatment trials, patients with cytogenetic abnormalities known to be associated with a poor prognosis are automatically assigned to intensive treatment arms or are excluded from the trial. Even for patients who are not treated in randomized trials, the alert clinician will take into account the cytogenetic findings when making a decision about what type of treatment to use. For example, a bone marrow transplant has inherent risks to the patient and is not recommended unless the risk of dying from the malignancy is substantially greater than the risk of undergoing a transplant.

   It has been supposed that the prognostic information derived from cytogenetic studies would be rendered irrelevant as treatment improved. In fact the improvements made so far have often tended to emphasize the prognostic differences, rather than diminish them. Furthermore, present forms of chemotherapy, including bone marrow transplantation, may not produce many more real "cures," however intense they become, and have deleterious side effects, including increasing morbidity. A cytogenetic result may therefore help the clinician to tailor the treatment to the needs of the patient, balancing the risk of relapse against the risk of therapy-related death or increased risk of a treatment-induced secondary malignancy. It would be unethical to give a patient with, for example, acute lymphoblastic leukemia and a good-prognosis chromosome abnormality the same desperate, intensive therapy as that called for if the patient had the Philadelphia translocation. It might also be unethical or unkind to impose intensive treatment on an elderly patient in whom chromosome abnormalities had been found that are associated with a very poor risk, when only supportive or palliative treatment might be preferred. There is a misconception that good-risk abnormalities such as t(8;21) are found only in young patients; the absolute incidence may be the same in all age groups ***(1)***. Therefore, older patients should not be denied access to a cytogenetic study that will help to ensure they are given treatment that is appropriate to their condition.
5. *To monitor response to treatment.* Conventional cytogenetic studies are not efficient for detecting low levels of clone, and therefore should not be used routinely to monitor remission status. FISH and molecular studies may be more appropriate. However, in the editor's laboratory, cytogenetic studies have detected a persistent clone in up to 12% of patients presumed to be in clinical remission

from leukemia, especially in those with persistent bone marrow hypoplasia (*unpublished observations*).

Some patients with chronic myeloid leukemia (CML) respond to interferon, and to the more recent therapy using STI 571; this response is usually monitored using six-monthly cytogenetic or FISH analysis.

It is sometimes helpful to confirm establishment of donor bone marrow after an allogeneic bone marrow or stem cell transplant, and this is easily done if the donor and recipient are of different sex. See the notes in chapter 12 about using cytogenetics in this context.

6. *To support further research.* Despite all that is already known, even in regard to the leukemias, there is still more to discover. Although the cytogeneticist in a routine laboratory may have little time available for pure research, there are ways that research can be supported. Publishing case reports, for example, brings information about unusual findings into the public domain. This makes it possible to collate the clinical features associated with such abnormalities, which leads to an understanding of the clinical implications, so helpful when the same abnormalities are subsequently discovered in other patients. Reporting unusual chromosome abnormalities can also indicate particular regions for detailed research analysis. For this reason, any spare fixed material of all interesting cases discovered should be archived in case it is needed. A less fashionable but no less important area of research is into the effect of secondary chromosome abnormalities. Some patients with "good-risk" abnormalities die and some with "poor-risk" abnormalities have long survivals; it is likely that knowledge of any secondary or coincident abnormalities present will help to dissect out the variations within good-risk and poor-risk groups *(2)*.

In the longer term, it is the hope that each patient will have a course of treatment that is precisely tailored to affect the malignant cells alone. Because the only difference between a patient's normal cells and malignant cells are the genetic rearrangements that allowed the malignancy to become established, it follows that such treatments will depend on knowing exactly what the genetic abnormalities are in each patient.

By the time that such treatment refinements become available, it is possible that conventional cytogenetic studies will have been replaced in some centers by emerging techniques such as microarrays. For the time being, however, a cytogenetic study remains an essential part of the diagnostic investigations of every patient who presents with a hematologic malignancy, and in many patients who present with certain solid tumors. This is not to deny the very valuable contributions made by other genetic assays, and the relative merits of these are compared in Chapter 17.

2.2. The Limitations of Conventional Cytogenetics Studies

A conventional cytogenetic study is still widely regarded as being the gold standard for genetic tests, since it is the best one currently available for assessing the whole karyotype at once. It is subject to limitations, however, including those described below. Where these can be overcome by using one of the new technologies, this is mentioned.

1. Only dividing cells can be assessed. This limitation is particularly evident in some conditions, such as chronic lymphocytic leukemia, malignant myeloma and many solid tumors, in which the available divisions, if any, may derive from the nonmalignant population. If it is already known (or suspected) what specific abnormality is present and there are suitable probes available, then some FISH and molecular analyses can be used to assess nondividing cells.
2. Analyses are expensive because of the lack of automation in sample processing and the time needed to analyze each division; consequently only a few divisions are analyzed. If available and applicable to the particular case, FISH and molecular analyzes have the advantage that hundreds or thousands of cells can be screened more efficiently.
3. There is no useful result from some patients; for example, if insufficient, unanalyzable, or only normal divisions are found. See Chapter 12 for a further consideration of the implications of finding only normal divisions. It is in the best interest of patients that the cytogeneticist seeks to minimize failures and to maximize clone detection.
4. Sometimes the abnormality found is of obscure significance. Rare or apparently unique abnormalities are still discovered even in well studied, common malignancies, and determining their clinical significance depends on a willingness to take the trouble to report them in the literature.

FISH and molecular analyses are generally used to detect known abnormalities, so the substantial proportion of unusual abnormalities that occurs is an argument in favor of retaining full conventional cytogenetic analysis for all cases of malignancy at diagnosis. It follows that these cases need to be published if the information gained is to be of any use to other patients.

5. The chromosome morphology may be inadequate to detect some abnormalities, or to define exactly what they are. In addition, some genetic rearrangements involve very subtle chromosome changes and some can be shown to happen through gene insertion in the absence of any gross structural chromosome rearrangement *(3)*. Such cryptic abnormalities are described in more detail in Chapters 3 and 5. A major advantage of FISH is that it can be used to unravel subtle, complex, and cryptic chromosome abnormalities.

References

1. Moorman, A. V., Roman, E., Willett, E. V., Dovey, G. J., Cartwright, R. A., and Morgan, G. J. (2001) Karyotype and age in acute myeloid leukemia: are they linked? *Cancer Genet. Cytogenet.* **126,** 155–161.
2. Rege, K., Swansbury, G. J., Atra, A. A., et al. (2001). Disease features in acute myeloid leukemia with t(8;21)(q22;q22). Influence of age, secondary karyotype abnormalities, CD19 status, and extramedullary leukemia on survival. *Leukemia Lymphoma* **40,** 67–77.
3. Hiorns, L. R., Min, T., Swansbury, G. J., Zelent, A., Dyer, M. J. S., and Catovsky, D. (1994) Interstitial insertion of retinoic receptor-α gene in acute promyelocytic leukemia with normal chromosomes 15 and 17. *Blood* **83,** 2946–2951.

2

Cytogenetic Studies in Hematologic Malignancies

An Overview

John Swansbury

1. The Challenge

The techniques for obtaining chromosomes from phytohemagglutinin (PHA)-stimulated lymphocytes for constitutional studies have been standardized to give consistent, reproducible results in almost all cases. It is therefore possible to refine and define a protocol that can be confidently used to provide an abundance of high-quality metaphases and prometaphases. For malignant cells, however, it can seem that every patient's chromosomes have an idiosyncratic reaction to the culture conditions, if the abnormal cells condescend to divide at all. For example, samples from different patients with leukemia can give widely different chromosome morphologies, even when processed simultaneously. In some cases it is also possible to recognize distinct populations of divisions on the same slide, often those with good morphology being apparently normal and those with poor morphology having some abnormality. It was once thought that poor morphology alone, even in the absence of detectable abnormality, might be sufficient to identify a malignant clone.

From: *Methods in Molecular Biology, vol. 220: Cancer Cytogenetics: Methods and Protocols*
Edited by: John Swansbury © Humana Press Inc., Totowa, NJ

However tempting this explanation has been to anyone who has seen such coexisting populations, such a hypothesis has not been subsequently confirmed. The formal demonstration of a clone in malignancy still requires the identification of some acquired genetic abnormality.

The high level of variation in chromosome quality associated with malignancy is often far greater than the improvements in quality that a cytogeneticist can make by altering the culturing and processing conditions, and by using different types of banding and staining. Some samples simply grow well and give good quality chromosome preparations, and others defy every trick and ruse in the cytogeneticist's repertoire, and produce small, ill defined, poorly spread, hardly banded, barely analyzable chromosomes.

Cytogenetic studies of malignancy therefore pose a particular technical challenge, and it is not possible to present a single technique that can be guaranteed to work consistently and reliably. In 1993 the author collated the techniques used for acute lymphoblastic leukemia by 20 cytogenetics laboratories in the United Kingdom, as part of a study for the U.K. Cancer Cytogenetics Group. Every step of the procedure was found to be subject to wide variation; the duration of exposure to hypotonic solution, for example, ranged from a few seconds to half an hour. It seemed that all permutations of technique worked for some cases, but no one technique worked consistently well for all cases.

Because the results are so unpredictable, every laboratory, and probably every cytogeneticist within each laboratory, has adopted a slightly different variation of the basic procedure. It is hard to demonstrate any real and consistent effect of these variations, and one suspects that some of them come and go with fashion, and others assume a mystical, almost ritual quality based more on superstition or tradition than on science. Furthermore, when a cytogeneticist moves from one laboratory to another, it often becomes evident that what worked well in one locality may not be effective in another, however faithfully the details are observed. For example, chromosome spreading has been shown to be affected by differences in atmospheric conditions ***(1)***, and in some places by differences in the water (whether distilled or deionized) used to make up the hypotonic potassium chloride solution (F. Ross, personal communication).

The techniques described in this book do work well in their authors' laboratories, and will work elsewhere; however, when putting them into practise in another laboratory, it may well be necessary to experiment with the details to determine what works best.

2. Type of Sample

2.1. Bone Marrow

For most hematology cytogenetics studies the vastly preferred tissue is bone marrow. Failures to produce a result can occur if the bone marrow sample is either very small or has an extremely high cell count. In either case, it is well worth asking for a heparinized blood sample as well.

One of the more significant factors in the overall improvement in success rates, abnormality rates, and chromosome morphology during the last two decades has been the better quality of samples being sent for analysis. This is a measure of the increasing importance that many clinicians now give to cytogenetic studies in malignancy. However, some clinicians do need to be persuaded to ensure that the sample sent is adequate. Apart from the fact that cytogenetic studies of bone marrow are expensive because they are so labor-intensive (and a great deal of time can be wasted on inadequate samples), more importantly, the once-only opportunity for a pretreatment study can be lost.

Ideally, a generous portion of the first spongy part of the biopsy should be sent, as later samples tend to be heavily contaminated with blood. Resiting the needle, through the same puncture if necessary, gives better results than trying to obtain more material from the same site. The sample must be heparinized; once a clot has started to form it will trap all the cells needed for a cytogenetic study. In Chapter 4, **Subheading 3.1.**, advice is given on how to attempt to rescue a clotted sample, but this is a problem better avoided than cured.

Usually 2 or 3 mL of good quality sample is sufficient; at least 5 mL may be needed if the marrow is hypocellular. However, it is the number and type of white cells present that is more important than the volume of the sample: each culture needs 1–10 million cells; several cultures need to be set up; most of the white blood cells in the peripheral circulation have differentiated beyond the ability to divide. If very little material is available, the whole syringe can be sent to the laboratory; any cells inside can then be washed out with some culture medium. Just one or two extra divisions can make the difference between success and failure. Conversely, if there is plenty of material and the laboratory has the resources, consider storing some of the sample as viable cells in liquid nitrogen, or as extracted DNA.

Heparinized bone marrow samples can be transported without medium if they will reach the laboratory within an hour or so. However, use of medium will reduce the likelihood of loss of material through clotting or drying, and the nutrients may help to preserve viability when the cell count is high.

The usual causes for a bone marrow sample being inadequate include (1) the patient is an infant, (2) the hematologist taking the sample is inexperienced, (3) the cell count is very low (especially in cases of myelodysplasia or aplastic anemia), or (4) the bone marrow has become fibrosed. Condition (4) produces what is often described as a "dry tap," as no bone marrow can be aspirated; in these circumstances, it can happen that production of blood cells (hemopoiesis) takes place in extramedullary sites (i.e., outside the bone marrow), such as the spleen. In some centers it is not regarded as ethical to request another bone marrow sample specifically for cytogenetic studies, probably because it is an unpleasant procedure for the patient. In other centers, however, a diagnostic cytogenetic study is regarded as sufficiently important to require a further aspirate if necessary. Standard culture conditions can be adapted to suit smaller samples ***(2)***, and Chapter 7 of this book has useful advice.

Although small or poor quality samples can sometimes fail to provide enough divisions for a complete study, it is the high-count samples that are most likely to fail completely. The vast majority of these cells are incapable of division, and their presence inhibits the few remaining cells that can divide. It is essential to set up multiple cultures and to ensure that the cultures do not have too many cells.

EDTA is not a suitable anticlotting reagent for cytogenetics studies and it should be declined in favor of heparin. However, if a sample arrives in EDTA, and there is no possibility of obtaining a heparinized sample, and the sample has not been in EDTA for long, then it is worth trying two washes in RPMI medium supplemented with serum and heparin before setting up cultures.

Sometimes the laboratory is offered cells that have been separated over Ficoll™ or Lymphoprep™. This process has an adverse effect on the mitotic index and such samples often fail. Washing twice in culture medium is sometimes helpful. If this is the only sample available, then fluorescence *in situ* hybridization (FISH) studies may have to be used instead of conventional cytogenetics.

2.2. Blood

Blood samples generally have a much higher failure rate and lower clone rate than bone marrow; also, the divisions may derive from cells that left the bone marrow some time previously, and so do not represent the current state of the disease. For all these reasons, blood samples may produce results that are more difficult to interpret. Therefore they should not be accepted willingly as an alternative to a good bone marrow sample, although they are better than nothing. It is sometimes said that a blood sample is worth studying only if there are blasts in the circulation; this may be true generally, but in the author's laboratory a clone has sometimes been detected even when no blasts have been scored.

2.3. Spleen

Occasionally a spleen biopsy is offered for cytogenetic studies of a patient with a hematologic malignancy. These generally work well enough: the biopsy should be washed in medium containing antibiotics, and minced with a sterile scalpel. The released cells are then treated as if they were from blood or bone marrow.

2.4. Solid tissues

For lymphomas and other solid tumors, a sample of the primary tissue is preferred. As described in Chapter 10, a clone may be found in a blood or bone marrow sample if it is infiltrated, and even occasionally in the absence of any signs of infiltration, but cytogenetic studies on these secondary tissues are an inefficient assay.

It occasionally happens that leukemic cells can accumulate to form a solid lump, such as a granuloma or chloroma, or can infiltrate the skin. Samples of such tissues may be sent to the cytogenetics laboratory for investigation. In general, they are best studied by FISH, especially if a previous bone marrow sample has already identified a clonal abnormality, but conventional cytogenetic studies are sometimes successful.

3. Common Causes of Failure

The preceding paragraphs have considered failure due to inherent limitations in the type of sample supplied. It can be frustrating for a laboratory to have to work with unsuitable or inadequate material, and any such deficiencies should be reported to the clinician. However, failures can also arise from errors in laboratory procedures, and every effort must be made to minimize these. Very often in cytogenetic studies of malignancy there is no possibility of getting a replacement sample: there may be only one biopsy taken, or only one bone marrow aspirated before treatment starts. Therefore it is wise to anticipate likely problems and try to avoid them. Chapter 12, **Subheading 4.** refers to quality control; having proper, documented procedures established for training, laboratory practical work, record-keeping, and so forth is essential both for ensuring that laboratory errors do not cause failures, and for detecting the cause of failures if they do occur.

If there suddenly seems to be a series of failures, then an immediate investigation must be started. However, every laboratory will have the occasional sample that fails, and sometimes there is no obvious reason. The following list may be helpful:

1. Contamination is usually obvious: cultures will be cloudy or muddy and may smell offensive; under the microscope the slides may show an obvious infestation with bacteria, yeast, or other microorganisms. If the contamination occurs only in particular types of culture, such as those stimulated with PHA or those blocked with fluorodeoxyuridine (FdUr), then it is likely that it came from this reagent.

   If all the cultures from one sample are infected, but those of another sample processed at the same time are not, then it is possible that the sample was contaminated at the source. In the author's experience, some clinicians have an unhelpfully casual attitude toward maintaining the sterility of samples.

   If there are any usable divisions on the slide, then it is likely that the infection arose late, possibly not during the culturing at all: it may have come from one of the reagents used in harvesting or banding.

   Procedures that will help to prevent contamination include steam sterilization of salt solutions, Millipore filter sterilization of heat-sensitive solutions, and the use of careful sterile technique when setting up cultures.
2. Check that the reagents have been made up correctly, being accurately diluted where appropriate. Errors in the reagents can be among the most difficult to detect; if this is suspected, it can be easier to discard all the reagents in current use and make up a fresh batch, rather than trying to track down exactly which one was at fault.
3. Check that the reagents have not deteriorated; many have a limited shelf life once they have been opened, and some need to be kept in the dark. It is often worthwhile to freeze small volumes and thaw one when needed. Once the reagent is thawed, do not refreeze, and discard any remainder after a week.
4. If the start of a series of failures coincides with the use of a new batch of medium or serum or some other reagent, a change of procedure, or the start of a new staff member, then this may be a clue to the source of the problem.
5. Check that the incubator is functioning properly, and had not overheated or cooled down.
6. Check that the types of culture set up were appropriate for the type of tissue or the diagnosis.

7. If there are no divisions at all, then possible reasons include: The tissue was incapable of producing any (as with most unstimulated blood cells), cell division was suppressed by exposure to extremes of heat or cold, the culture medium was unsuitable for supporting cell growth (e.g., because of a change of pH), too many cells were added to the culture, the arresting agent (colcemid or colchicine) was ineffective, all the dividing cells had been lysed by too long exposure to hypotonic solution, or that all the chromosomes had been digested off the slide by too long exposure to trypsin.
8. If there are divisions but the chromosomes are too short, then possible reasons include the addition of too much arresting agent, or too long an exposure to the arresting agent. Short chromosomes can also be a feature of the disease—the chromosomes from a high hyperdiploid clone in acute lymphoblastic leukemia (ALL) can be very short in some cases, despite every effort to obtain longer ones.
9. If the chromosomes are long and overlapping, and arranged in a circle with the centromeres pointing toward the center (this is known as an anaphase ring), then the concentration of arresting agent was too low to destroy the spindle.
10. If the chromosomes have not spread and are too clumped together, then possible causes include ineffective hypotonic solution, too short an exposure to hypotonic solution, or poor spreading technique—if the slide was allowed to dry too quickly after dropping the cell suspension onto it, then the chromosomes will not have chance to spread out. However, if the chromosomes are also fuzzy, then it is also possible that their poor quality is intrinsic to their being malignant. Such cases will tend to produce poor chromosomes whatever technique is tried, and there is little that can be done about them.
11. If the chromosomes are not analyzable owing to lack of a clear banding pattern then this is usually attributable to a combination of how old the preparations were before banding and how long they were exposed to trypsin. Slides can be aged at room temperature for a week, for a few hours in an oven, or for a few minutes in a microwave, but this is an essential step before banding is effective.

4. Time in Transit

The samples should be sent to the laboratory as quickly as possible without exposure to extremes of temperature. A result can sometimes be obtained even from samples a few days old, with myeloid disorders being generally more tolerant of delay. Samples from lymphoid disorders, however, and all samples with a high white blood cell count, usually need prompt attention. If there is plenty of culture medium, some samples can survive for 2 or 3 d, preferably kept at a cool temperature but not below 4°C. In such circumstances, extra cultures should be set up once the sample arrives, giving some of them 24 h in the incubator to recover before starting any harvesting. However, the chances of failure increase rapidly with increasing delay in transit.

5. Safe Handling of Samples

All samples should be handled as carefully as if they might be contaminated with hepatitis virus or HIV (AIDS). Suitable laboratory protective clothing (including coats/aprons and gloves) should be worn. Plastic pipets or "quills" should be used (rather than needles or glass pipets) while processing unfixed tissue, to avoid the risk of needlestick injury.

It is possible to use just a clear, draft-free bench for all cytogenetics laboratory work. However, it is greatly preferable to use a laminar flow cabinet for all processing and handling of unfixed samples, with a vertical flow of air to protect both the sample from contamination and the cytogeneticist from infection.

Low levels of sample contamination are not usually a problem, as the medium contains antibiotics and most cultures are short term. However, it is good practice always to use careful sterile technique. Pipets and culture tubes must be sterile. Disposable plastic tubes are most convenient; reusable glass tubes can be used for cultures and processing, but should be coated with silicone (e.g., using dimethyldichlorosilane, in 1,1,1-trichloroethane), as otherwise all the divisions will stick to the inside of the glass as soon as they are fixed.

The risk to the cytogeneticist of infection from aerosols derived from marrow or blood is low except during centrifugation, when closed containers must be used. Most centrifuges blow air around the rotor to keep it cool during operation.

Once the sample is fixed, it poses no risk; however, be aware that the outside of the tube may still be contaminated. At the end of the work, all flasks, tubes, pipets, gloves, tissues, and so forth that have been (or which *could* have been) used for sample processing must be discarded into an appropriate container and treated separately from "clean" waste such as paper.

Many of the reagents used in the cytogenetics laboratory are harmful or potentially harmful; the laboratory should provide all its staff with appropriate advice on the safe use and disposal of these, and what to do in the event of a spillage or accident.

6. Choice of Cultures in Hematology Cytogenetics

The duration of the malignant cell cycle varies greatly between patients: a range of 16–292 h was obtained in a series of 37 patients with acute myeloid leukemia (AML) *(3)*. There appear to be no obvious indicators of what the cycle time might be for a patient, so it is not possible to predict exactly which culture will give the best result. Therefore one of the most significant factors in getting a successful result is the setting up of multiple cultures to maximize the chances of getting abnormal divisions. Different cell types tend to come into division after different culture times, so, depending on the diagnosis, certain cultures are more likely to have clonal cells than others *(4,5)*. This has been taken into account for the cultures that are recommended in the following chapters. However, extra cultures should always be set up when materials and manpower permit. The different culture types are describe in the following subheadings.

6.1. Immediate Preparation

This type of preparation is also known as "direct" in some laboratories (*see* Chapter 7). As soon as the sample is aspirated from the patient, two drops are put straight into a solution of warmed, hypotonic KCl that also contains colcemid and heparin *(6)*, and 10% trypsin *(7)*. Twenty-five minutes later the tube is centrifuged and fixed according to the usual procedures.

This technique has been said to give high success rates and clone detection rates. However, in most centers it is not possible to organize such close cooperation between clinic and laboratory.

6.2. Direct Preparation

The sample is harvested the day it was taken. Colcemid may be added immediately when setting up cultures or after an hour or so of incubation. Harvesting usually begins about an hour after colcemid is added. This type of culture is not suitable for most types of AML, in which it usually produces only normal divisions.

6.3. Overnight Culture

Colcemid is added to the culture at the end of the afternoon; the culture is then incubated overnight and harvested the next morning. This is the culture most likely to produce some divisions if the overall mitotic index in the sample is low. Colcemid arrests cell division by preventing spindle formation during mitosis, and so the chromatids cannot separate. The longer the colcemid is left in the culture, the more divisions are accumulated but the shorter the chromosomes become. Most divisions in an overnight culture will probably have short chromosomes but often there are some with chromosomes long enough to be analyzable. This type of culture has sometimes been described as producing "hypermetaphase" spreads, when large numbers of divisions are needed but chromosome quality is not so important, as in FISH studies.

Some centers include an element of synchronization by putting the culture into the refrigerator (at not less than 4°C) until about 5 P.M. before being put into the incubator overnight, then starting the harvest at about 9 A.M. next morning. Because samples often cool down between collection and arrival in the laboratory, deliberately putting them into the refrigerator introduces a way of controlling the recovery. Although it is not possible to predict precisely when the cells in any particular sample will start to divide again after the temperature is restored, it has been determined that in many cases it is about 14.25 h for chronic myeloid leukemias (CMLs) and 16.25 h for other disorders *(8)*.

6.4. Short-term Cultures

The sample is incubated for one, two, or three nights before harvesting. Culturing for just one night is regarded as giving the highest overall clone detection rates in leukemias, especially in myeloid disorders.

6.5. Blocked Cultures (Synchronization)

The divisions are probably not truly synchronized, the effect arising through a retarding of the S-phase; "blocking" is therefore a better term. These methods were introduced to increase the number of divisions collected with a short exposure to colcemid, thus obtaining long chromosomes *(9)*. In practice, the number of divisions obtained in malignancy studies is usually reduced, or there may be none at all. The duration of the mitotic cycle of leukemic cells (and therefore the release time) is more variable, and usually considerably longer, than that of normal tissues. A short exposure to colcemid is usually used (but see the variation described in Chapter 4), which means that there is a strong chance of missing the peak of divisions when it happens. However, if this method does work, it can give good quality chromosomes, so it is always worth doing if there is sufficient material.

Commonly used synchronizing agents are methotrexate (Amethopterin) *(10)*, fluorodeoxyuridine *(11)* and excess thymidine *(1)*. The first two tend to be better for myeloid disorders, with the last being better for lymphoid disorders.

These published studies reported that the release time should be 9.5–11.5 h for myeloid and leukemic cells *(9)*, and that that the time varies between patients, and showed that the cell cycle time is generally shorter in CML than in AML *(10)*. Despite this, many laboratories routinely allow only 4 or 5 h of release before adding colcemid.

6.6. Mitogen-Stimulated Cultures

Mature lymphocytes do not divide spontaneously, but will transform (become capable of division) as part of their immune response. Certain reagents, termed mitogens, are regularly used in cytogenetics studies to stimulate lymphocytes into division, and some of these are described in Chapter 9. However, the disease may affect lymphoid cells so that they are not capable of responding to mitogens, or the treatment may suppress the immune response; in these cases mitogens will not be effective in producing divisions.

If the lymphocytes have already been transformed, for example, because the patient has an infection, then lymphocyte divisions can be found in unstimulated cultures. Immature lymphocytes that are still dividing do not usually enter the circulation and are rare in the normal, healthy state, but can be common in hematologic malignancy when the bone marrow organization is in disorder.

References

1. Wheater, R. F. and Roberts, S. H. (1987) An improved lymphocyte culture technique: deoxycytidine release of a thymidine block and use of a constant humidity chamber for slide making. *J. Med. Genet.*, **24,** 113–115
2. Brigaudeau, C., Gachard, N., Clay, D., Fixe, P., Rouzier, E., and Praloran, V. (1996) A 'miniaturized' method for the karyotypic analysis of bone marrow or blood samples in hematological malignancies. *Pathology* **38,** 275–277.
3. Raza, A., Maheshwari, Y., and Preisler, H. D. (1987) Differences in cell characteristics among patients with acute nonlymphocytic leukemia. *Blood* **69,** 1647–1653.
4. Berger, R., Bernheim, A., Daniel, M. T., Valensi, F., and Flandrin, G. (1983) Cytological types of mitoses and chromosome abnormalities in acute leukemia. *Leukemia Res.* **7,** 221–235.
5. Keinanen, M., Knuutila, S., Bloomfield, C. D., Elonen, E., and de la Chapelle, A. (1986) The proportion of mitoses in different cell lineages changes during short-term culture of normal human bone marrow. *Blood* **67,** 1240–1243.

6. Shiloh, Y. and Cohen, M. M. (1978) An improved technique of preparing bone-marrow specimens for cytogenetic analysis. *In Vitro* **14,** 510–515
7. Hozier, J. C. and Lindquist, L. L. (1980) Banded karyotypes from bone marrow: a clinical useful approach. *Hum. Genet.* **53,** 205–9.
8. Boucher, B. and Norman, C. S. (1980) Cold synchronization for the study of peripheral blood and bone marrow chromosomes in leukemias and other hematologic disease states. *Hum. Genet.* **54,** 207–211
9. Gallo, J. H., Ordonez, J. V., Grown, G. E., and Testa, J. R. (1984) Synchronisation of human leukemic cells: relevance for high-resolution banding. *Hum. Genet.* **66,** 220–224.
10. Morris, C. M., and Fitzgerald, P. H. (1985) An evaluation of high resolution chromosome banding of hematologic cells by methotrexate synchronisation and thymidine release. *Cancer Genet. Cytogenet.* **14,** 275–284.
11. Webber, L. M. and Garson, O. M. (1983) Fluorodeoxyuridine synchronisation of bone marrow cultures. *Cancer Genet. Cytogenet.* **8,** 123–132.

3

The Myeloid Disorders

Background

John Swansbury

1. Introduction

Malignant myeloid disorders have broadly similar responses to cytogenetic techniques and many have similar chromosome abnormalities. Included are diseases that are frankly malignant, such as acute myeloid leukemia (AML), and some that may be regarded as premalignant, such as the myeloproliferative disorders (MPD). A proportion of the premalignant group may progress to acute leukemia but they are serious diseases in their own right, often difficult to treat, and may be fatal. They are all clonal disorders, that is, the bone marrow includes a population of cells ultimately derived from a single abnormal cell, which usually tends to expand and eventually suppress or replace the growth and development of normal blood cells. This group of disorders includes the following:

The myeloproliferative disorders (MPD)
The chronic myeloid leukemias (CML)
The myelodysplastic syndromes (MDS)
Aplastic anemia (AA)
Acute myeloid leukemia (AML)

From: *Methods in Molecular Biology, vol. 220: Cancer Cytogenetics: Methods and Protocols*
Edited by: John Swansbury © Humana Press Inc., Totowa, NJ

The major clinical and cytogenetic features of the myeloid malignancies are summarized in the following subheadings.

2. The Myeloproliferative Disorders

In general terms, the MPDs have too many of one kind of myeloid cell. In many cases the disease is chronic, slowly evolving, and the symptoms can be controlled for many years with relatively mild cytotoxic treatment. However, they are serious diseases and a true cure is difficult to obtain. Although they are clonal disorders, the incidence of chromosomally identified clones is low except for chronic granulocytic leukemia (CGL, *see* **Subheading 2.4.**). This may be because the cells with abnormal chromosomes are in too low a proportion to be detected by a conventional cytogenetic study (in which only 25 divisions may be analyzed). Alternatively, visible chromosome rearrangements may be late events in the course of the disease; their occurrence may be necessary for the disease to progress to more severe stages, culminating in AML in some cases. AML secondary to MPD or MDS tends to be refractory to treatment: cytotoxic chemotherapy often fails to eradicate the clone and usually results in prolonged myelosuppression with poor restoration of blood counts. This may be because the prolonged antecedent disorder has compromised the ability of normal myeloid cells to repopulate the marrow. In CGL, disease progression is inevitable and is referred to as blast crisis.

2.1. Polycythemia Rubra Vera

Polycythemia rubra vera (PRV) is an excess of red blood cells. The incidence of detected cytogenetic clones is low, about 15%. The abnormalities found include those seen in all myeloid disorders but with deletion of the long arms of chromosome 20 being most common. There are two forms of this abnormality: del(20)(q11q13.1) and the smaller del(20)(q11q13.3) ***(1)***.

Treatments for PRV include venesection to reduce the load of red cells and the use of radioactive phosphorus (^{32}P) or busulfan to suppress the production of red cells. The cytotoxic treatments do carry a small risk of promoting a progression from premalignancy to malignancy, or the development of secondary malignancy.

2.2. Essential Thrombocythemia (ET)

Essential thrombocythemia (ET) is an excess of and/or abnormal platelets. This is a rare condition, and using conventional cytogenetics studies, no clone is found in most patients; in one large series only 29/170 (5%) of cases had a clone ***(2)***. The most commonly reported abnormality is the Philadelphia translocation, t(9;22)(q34;q11), and this has been detected by fluorescence *in situ* hybridization (FISH) testing positive for *BCR/ABL* in as many as 48% of cases ***(3,4)***. However, other authors have not been able to detect *BCR/ABL* in their patients ***(5,6)***. Clearly, there are as yet unresolved issues about the precise diagnosis of ET, and about the relationship between ET and CGL. For practical purposes, the cytogeneticist needs to be aware that discovering a t(9;22)(q34;q11) by cytogenetics or a *BCR/ABL* rearrangement by FISH in a patient with a diagnosis of ET does not necessarily mean that the diagnosis must be changed to CGL.

2.3. Myelofibrosis and Agnogenic Myeloid Metaplasia

The bone marrow is replaced by fibrous tissue and blood cell production may take place in extramedullary sites (outside the bone marrow) such as the spleen, which causes the spleen to enlarge. Deletion of part of the long arms of a chromosome 13 is common, as is a dicentric chromosome dic(1;7)(q10;p10), which results in gain of an extra copy of the long arms of chromosome 1 and loss of the long arms of a chromosome 7. This abnormal chromosome is similar in appearance to a normal chromosome 7, and can be missed by an inexperienced cytogeneticist.

2.4. Chronic myeloid Leukemia and Chronic Granulocytic Leukemia

CML is often taken to be synonymous with CGL, but actually also includes rarer disorders such as the chronic neutrophilic, eosinophilic, and basophilic leukemias, juvenile chronic myeloid leukemia; and chronic myelomonocytic leukemia (*see* **Subheading 2.**). In all there is an excess of white blood cells. CGL is often considered in its own right, rather than as part of the MPD group, as it has a distinct cytogenetic and clinical character. In more than 90% of cases the Philadelphia translocation (abbreviated to Ph) is present, usually as a simple translocation between chromosomes 9 and 22, t(9;22)(q34.1;q11) (Fig. 1). In about half of the remaining cases, called Ph-negative CGLs, it can be shown by molecular methods that the same genes (*ABL* and *BCR*) are rearranged even though the chromosomes appear normal.

The natural history of CGL is of a relatively mild chronic phase that is followed by disease acceleration into an acute phase known as blast crisis. The chronic phase is of variable duration; it may be over before the patient is first diagnosed, and it can last for 15 yr or more. The stimulus for acceleration is at present unknown. In some patients, chronic phase bone marrow can be harvested and stored for use as an autograft at a later stage. Although this procedure can restore the patient to chronic phase disease, it tends to be of shorter duration. It has been found that treatment with interferon increases the number of Ph-negative divisions in some patients, and a few have become entirely hematologically normal, although probably not cured. More recent treatments that have a greater effect, such as STI 571 (Gleevec™), may have a wider application.

It is useful to have a cytogenetic study at diagnosis, against which to compare the results of subsequent studies. There has not been agreement about the prognostic effect of secondary abnormalities identified at diagnosis, but most of them are not thought to be adverse clinical signs *(7)*. Some abnormalities, such as trisomy 8 and gain of an extra der(22), have been associated with a poorer prognosis. However, if secondary abnormalities are detected during the course of the disease, then this is a stronger indication that acceleration of the disease is imminent. Cytogenetic studies of large numbers of divisions have shown that in some cases these late-appearing abnormalities were present at diagnosis, but at a very low incidence (B. Reeves, *unpublished observations*). The introduction of FISH

Inversion 3
inv(3)(q21q26)

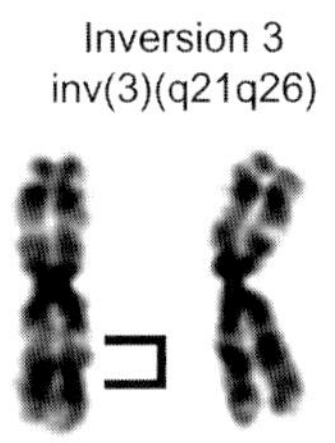

The normal #3 is on the left. This abnormality is one that can be missed by the inexperienced, especially when other abnormalities are also present. It is readily recognized by the dark band midway down the long arms.

t(6;9)(p21;q34.3)

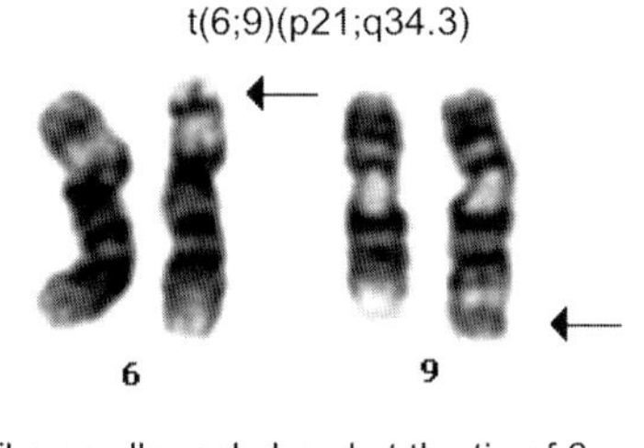

The smaller pale band at the tip of 6p can be subtle. However, the 9q+ is obvious and is usually slightly darker than the 9q+ arising from the Philadelphia translocation, shown below.

t(9;11)(p21-22;q23)

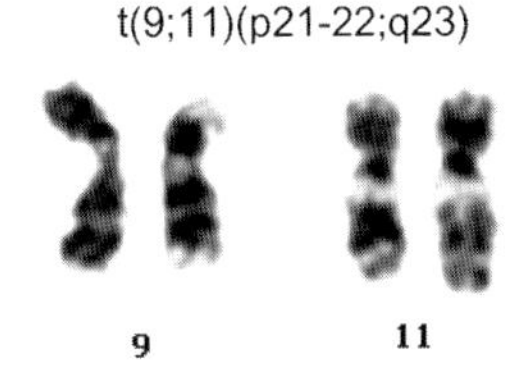

Many of the translocations involving 11q23 are subtle and can easily be missed. Here the abnormal chromosomes are on the right of each pair, showing a paler 9p and darker 11q.

t(9;22)(q34.1;q11)

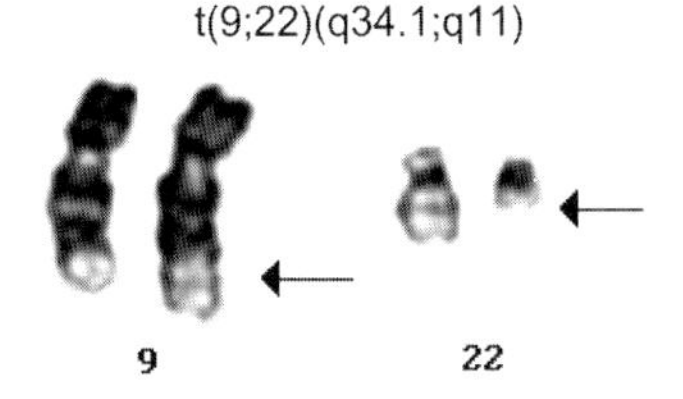

The Philadelphia translocation, in its simple form, soon becomes familiar. However, as with most of the abnormalities illustrated here, it can be masked by involvement with other chromosomes.

t(15;17)(q24;q21)

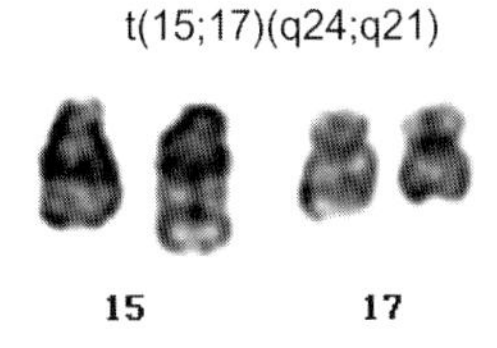

The normal chromosome is on the left of each pair. The 15q+ is not often as obvious as in this case; frequently it resembles a #14. Similarly, in poorer-quality metaphases, the 17q- can resemble a #20.

Inversion 16
inv(16)(p13q22)

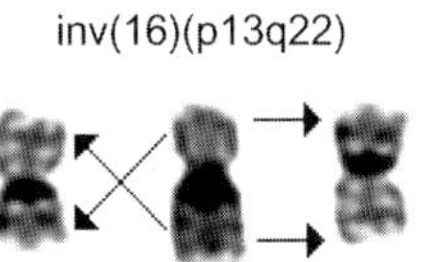

The chromosome in the center is the normal 16. Notice that in this person it has a larger dark heterochromatic region below the centromere. The same inverted 16 is shown on either side: on the left the centromere is in the correct orientation; on the right the telomeres are in the correct location.

Fig. 1. Examples of recurrent abnormalities in myeloid disorders, particularly illustrating some that can be subtle.

analysis using probes for the *ABL* and *BCR* genes led to the discovery that approx 10% of translocations include deletion of part of one of these genes, usually the proximal part of ABL, and this has been strongly associated with a poor prognosis ***(8)***.

Many recurrent secondary chromosome abnormalities are seen in CGL, and these tend to accumulate in major and minor pathways ***(9)***. The major abnormalities are +8, +19, +der(22), and i17q. Some abnormalities are associated with distinct types of blast crisis. For example, the isochromosome for the long arms of a chromosome 17 (now known to be a dicentric chromosome with breakpoints at 17p11) ***(10)*** is associated with myeloid blast crisis, and abnormalities of 3q21 and/or 3q26 (Fig. 1) are associated with megakaryocytic blast crisis.

It can be difficult to distinguish clinically between Ph+ acute lymphoblastic leukemia (ALL) and CGL presenting in lymphoid blast crisis. A molecular study of the *BCR/ABL* fusion gene product can help, since almost all CGLs have a 210-Kda product, whereas about 50% of ALLs have a 190-Kda product. The presence of normal divisions found by a conventional cytogenetic study is sometimes helpful, as most CGLs have only one or two, and some ALLs have a higher proportion. However, a cytogenetic study of a bone marrow sample taken after starting treatment provides further evidence: In CGLs, the Ph persists throughout chronic phase, but in ALLs it usually disappears once the disease is in remission.

3. The Myelodysplastic Syndromes

Historically there have been many terms for these disorders, including dysmyelopoietic syndrome, preleukemia, subacute leukemia, and smouldering leukemia. Transformation into acute leukemia does occur, but these are not merely preleukemic conditions; they are malignant, clonal diseases in their own right. They have abnormal growth (dysplasia) or failure of maturation of one or more cell lineages in the bone marrow, usually resulting in a deficiency of one or more blood components. For example, dyserythropoiesis indicates abnormalities of the cells that produce erythrocytes (red blood cells), which results in anemia. All three lineages may be involved (trilineage dysplasia), leading to pancytopenia (inadequate numbers of all blood elements: red cells, white cells, and platelets). MDS was primarily divided into subgroups according to an arbitrary but generally useful scheme based on the percentage of blast cells in the bone marrow: (1) Refractory anemia (RA), which had up to 5% blasts; (2) RAEB (RA with excess of blasts) had up to 20%; and (3) RAEBt (RAEB in transformation) which had up to 29% ***(11)***. Blasts amounting to 30% or more was said to define acute leukemia. Various other disease types were also classed as MDS, including RARS (refractory anemia with ring sideroblasts); chronic myelomonocytic leukemia (CMML); the 5q- syndrome ***(12)***, which is a relatively mild, indolent condition that has the longest median survival of any class of MDS; and juvenile monosomy 7 syndrome ***(13)***, which is associated with a poor prognosis.

However, this well established classification has recently been modified by the World Health Organization (WHO), and is now as follows:

1. Refractory anemia ± sideroblasts: < 10% dysplastic granulocytes.
2. Cytopenia: May have bilineage or trilineage dysplasia but < 5% blasts.
3. RAEB 1: With 5–10% blasts.
4. RAEB 2: With 11–19% blasts.
5. CMML in either MDS or MPD.
6. 5q-syndrome.

Note that the RAEBt class has been abolished, such that the presence of 20% blasts now defines acute leukemia. Like the MPDs, most of the MDSs are usually slow-evolving disorders in which supportive treatment may be adequate in the early stages; aggressive cytotoxic treatment rarely produces a remission and is more likely to induce bone marrow failure or acceleration of disease progression. The risk of developing acute leukemia (usually AML) increases in each subtype of MDS, but many patients eventually die of the consequences of marrow failure associated with MDS without progressing to overt leukemia.

It is important to distinguish MDS from similar clinical conditions that are not clonal, as many of the signs of MDS can also occur in nonmalignant disorders. Anemia is one of the most common clinical signs of MDS, but in most cases anemia has a benign cause and responds to treatment with supplements such as iron or folic acid (vitamin B_{12}). It may also be a side effect of treatment for other disorders, such as lithium for depression. In particular, chemotherapy for some other malignancy usually has a profound effect on the bone marrow, and in some cases it can be difficult to distinguish between a reaction to chemotherapy and an MDS which, as a new, secondary malignancy, may have been caused by that chemotherapy.

In all these areas of diagnostic uncertainty, cytogenetic studies can help: If a chromosomally abnormal clone is found, this is very strong evidence that the condition is malignant. The incidence of clonal chromosome abnormalities increases with each subtype, from as low as 10% up to nearly 50%. Failure to find a clone may not mean that there is no cytogenetically abnormal clone present, but rather that it may be at too low a level to be detected by a conventional cytogenetic study.

In MDS, as in other hemopoietic diseases, some cytogenetic abnormalities are associated with a poor prognosis (e.g., complex clones that include loss or deletion of part of the long arms of chromosomes 5 and/or 7) and some can indicate a relatively benign course (e.g., deletion of part of the long arms of a chromosome 5 as the sole cytogenetic abnormality as part of a "5q- syndrome" ***(12)***. Most of the chromosome abnormalities found in AML also occur in MDS, but some specific translocations are found rarely or not at all; these include t(8;21)(q22;q22), t(15;17)(q24;q21), and inv(16)(p13q22). The latest WHO classification of MDS defines as AML any disease having these translocations even if the number of blasts is < 20%.

CMML is identified by an absolute monocyte count of $> 2 \times 10^9$/L. The number of blasts is variable and is not used to define or subdivide this category. This is unfortunate; because the number of blasts correlates with prognosis, it follows that the overall survival for all types of CMML combined is intermediate. A clone is found in about

25–30% of cases. Although there is no common characteristic chromosome abnormality associated with CMML, there are several recurrent but rare abnormalities. These include translocations involving 5q33 (e.g., t(5;12)(q33;p13), associated with eosinophilia) and 8p11-12 ***(14)***, which is associated with a syndrome having an acute phase of T-lymphoblastic lymphoma; the most common translocations are t(8;13)(p11;q12), t(8;9)(p11;q32), and t(6;8)(q27;p11).

4. Aplastic Anemia

AA is a condition in which there may be almost complete absence of blood-forming tissue in the bone marrow. There are three main causes: (1) It may be secondary to a major exposure incident, for example, radiation or poisoning with benzene. (2) AA is also associated with a congenital condition, Fanconi anemia. These patients have a defect in DNA repair, which is often evident by the large number of random breaks and gaps seen in chromosomes, especially when grown in low-folate medium. Approximately 10% of patients with of Fanconi anemia will develop MDS or AML. (3) AA also occurs without known cause, and in at least some cases a clonal cytogenetic abnormality can be detected. Because there are usually very few cells in the sample sent to the cytogenetics laboratory, it is a difficult disease for cytogenetic study. The most commonly found abnormalities are those also seen in other myeloid malignancies, such as 5q-, –7, and +8, which is evidence that in these cases the AA is a form of MDS ***(15)***. However, trisomy 6 is a recurrent finding in AA that is rare MDS and AML ***(16)***.

5. Acute Myeloid Leukemia

There are eight FAB (French–American–British) classification ***(17,18)*** types of AML, some of which are subdivided further. All the chromosome abnormalities that occur in MDS and MPD also occur in AML, although in different proportions. However, there are some abnormalities that occur in AML that are extremely rare in other disorders, including t(8;21)(q22;q22), t(15;17)(q24;q21), and inv(16)(p13q22). It may be no coincidence that these abnormalities are generally confined to granulocytic cells and are associated with a good prognosis, while most other abnormalities tend to occur in all kinds of myeloid cells and are broadly associated with a poorer prognosis.

5.1. Cytogenetic Abnormalities with Strong AML FAB-Type Associations

M1: Myeloblastic leukemia without maturation of the blast cells; there is no specific cytogenetic abnormality, although trisomy 13 is most common in M0 and M1 ***(19)***. It is associated with a poor prognosis.

M2: Myeloblastic leukemia with maturation; the most common abnormality is t(8;21)(q22;q22). As previously mentioned, occasional cases with t(8;21) were said to have a diagnosis of MDS; in some of these, it has been found that the precise number of blast cells present was uncertain because of ambiguous morphology, and so the diagnosis could have been AML. However, all cases with a t(821) are now defined as having AML, however low the blast count may be.

The t(8;21) is associated with a high remission rate, and consequently a relatively good prognosis for AML. However, there were very few long-term survivors before the introduction of modern intensive chemotherapy.

A very common abnormality secondary to t(8;21) is loss of an X chromosome in female patients or the Y chromosome in males. Loss of a sex chromosome is very rare in AML except in the presence of a t(8;21), so it clearly has a specific role in this situation, one that is at present unknown. Another common secondary abnormality is deletion of part of the long arms of chromosome 9. This has been found as the sole event in some cases of AML, and it was suggested that it may indicate the presence of a cryptic t(8;21). However, FISH and molecular studies have shown that this was usually not present ***(20)***.

Although they are so closely associated with t(8;21), the clinical significance of these secondary abnormalities is not known. Several published series have reported contradictory effects on prognosis ***(21)***. Although t(8;21) is used to identify a good-risk group in AML ***(23)***, some patients do not respond well to treatment and it would be of great help to the clinician to be able to distinguish these patients from those who will do well.

Molecular evidence of persistence of t(8;21) has been found in some patients more than 7 yr in remission, with no evidence for tendency to relapse ***(24)***.

M3 & M3v: Promyelocytic leukemia. This is characterized by a t(15;17)(q24;q21) (Fig. 1), a highly specific abnormality that is found elsewhere only in a rare form of CGL promyelocytic blast crisis. Clinical features include disseminated intravascular coagulation (DIC), a life-threatening condition that is the cause of many early deaths in M3. Once this crisis has passed, the prognosis for the patient is good. In particular, the leukemic cells respond to all-*trans*-retinoic acid (ATRA) by proceeding with differentiation and normal apoptosis, so this is used as part of the treatment. The quoted breakpoints on chromosomes 15q and 17q vary widely among different publications; the author favors those proposed by Stock et al. ***(22)***.

The effect of the presence of secondary abnormalities is uncertain. In one study ***(23)*** (in which all secondary abnormalities were combined) they appeared to have no effect, but in others ***(25,26)*** the co-occurrence of trisomy 8 reduced the prognosis from good to standard. It would seem reasonable to expect that different secondary abnormalities have a different effect on prognosis.

Unlike the case with t(8;21), the detection of t(15;17) in remission is usually a sign of imminent relapse. Because the chromosome quality of t(15;17)+ cells is often poor, and the abnormality is difficult to see with poor-quality chromosomes (Fig. 2), FISH should be used for follow-up studies using the probes that are available for the PML (at 15q24) and retinoic acid receptor alpha (RARA) at 17q21 gene loci. Molecular methods appear to be too sensitive for clinical use at present, as they detect residual disease in more patients than those who proceed to relapse ***(27)***.

Another translocation involving the same gene on chromosome 17 plus the *PLZF* gene at 11q23 is the t(11;17)(q23;q21), which can

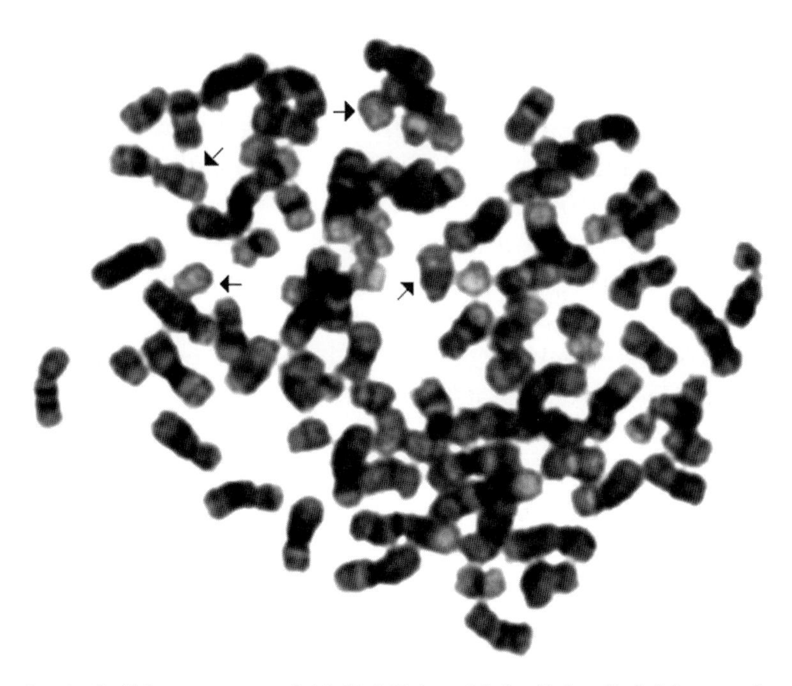

Fig. 2. Cell from a case of AML M3 in which all the diploid metaphases found were normal and all the tetraploid metaphases were too poor for full analysis. However, the typical t(15;17)(q24;q21) could still be recognized; the abnormal chromosomes are indicated with *arrows*.

also occur with a diagnosis of M3 ***(28)***. However, these patients do not respond in the same way to ATRA. A cytogenetically identical t(11;17)(q23;q21) is also found in AML M5, but the genes involved are *MLL* and *AF17*.

M4: Myelomonocytic leukemia; t(8;21)(q22;q22) also occurs, although at a lower frequency than in M2. A well characterized subtype, M4eo (M4 with abnormal eosinophilia), is strongly associated with inv(16)(p13q22) (Fig. 1) and the rarer t(16;16)(p13;q22). This abnormality has been associated with a relatively good prognosis, although with a tendency to central nervous system relapse. The inversion is not easy to identify in poor quality chromosomes, especially because the heterochromatic region of chromosome 16 varies considerably in size. A common secondary abnormality is trisomy

22, so if this is seen the 16s should be carefully checked. If there is any doubt, a FISH study will determine whether or not an inv(16) is present. There have been conflicting reports as to whether or not a trisomy 22 as the sole abnormality is likely to indicate the presence of a cryptic inv(16) ***(20,29)***.

A del(16)(q22) is also a recurrent abnormality in myeloid malignancy; the interpretation of the significance of this abnormality requires more care, as in M4eo it is probably a variant of the inv(16) or t(16;16) and may indicate the same good prognosis; but in other conditions, such as MDS, it has been associated with a poor prognosis ***(30)***.

M5: A t(8;16)(p11;p13) occurs in both M4 and M5. This abnormality is also linked with other clinical features, including disturbance of clotting function ***(31)***, which can mimic the DIC found in M3, but it is particularly associated with phagocytosis. Genes located at 8p11 are also involved in translocations with many other chromosomes ***(14,32)***, which seem to specify the type of malignancy produced.

M5 is divided into two FAB subtypes:

M5a (monoblastic leukemia) is generally associated with t(9;11)(p21-22;q23). This is a subtle abnormality and can be missed unless the 9p and 11q regions are specifically checked (Fig. 1). In the author's laboratory, a study using a FISH probe for the *MLL* gene in a series of patients identified one with a t(9;11) that had been missed ***(33)***. Other translocations involving *MLL* at 11q23 also tend to be more common in M5a.

M5b (monocytic leukemia) is not closely associated with any particular cytogenetic abnormality.

M6: Erythroleukemia: no specific cytogenetic abnormality, but about 25% of all occurrences of t(3;5)(q21-25;q31-35) are found in M6.

M7: Megakaryocytic leukemia; abnormalities of 3q21 and/or 3q26 are more common. People with Down syndrome (constitutional trisomy 21) have an increased risk of developing leukemia, and often this is of the M7 type. A highly specific abnormality, t(1;22)(p22;q13), is associated with M7 in infants ***(34,35)***.

5.2. Cytogenetic Abnormalities in AML Without FAB-Type Associations

As well as the AML-associated cytogenetic abnormalities already mentioned, which show some degree of FAB-type specificity, there are others that do not. Of these, trisomy 8 is the only one that is found in M3/M3v; the others occur in FAB types except for M3.

Abnormalities of chromosomes 5 and 7 usually take the form of loss of the whole chromosome or deletion of part of the long arms. In most cases other chromosome abnormalities are also present, and the prognosis is generally poor. These abnormalities are particularly common in MDS and AML that are secondary to exposure or to treatment for other malignancies that commenced at least 2 yr previously.

Trisomy 8 is the most common abnormality in AML, occurring both alone and in combination with other abnormalities. The prognosis is generally regarded as being intermediate or poor, and it has been claimed that the prognosis depends on what other abnormalities are present ***(36)***. If the chromosome morphology is poor, trisomy 10 (a rare finding but one that may indicate a poorer prognosis) may be missed on the presumption that it is the more common trisomy 8.

The Philadelphia translocation, t(9;22)(q34;q11), occurs in about 3% of AML cases, and is associated with a poor prognosis.

As previously mentioned, abnormalities of bands 3q21 and 3q26 are very frequently associated with dysmegakaryopoiesis; these abnormalities have been found in various hematologic disorders and generally indicate a poor prognosis ***(37)***.

Lastly, a specific translocation, t(6;9)(p23;q34.3), is associated with AML that is TdT+ (i.e., expresses terminal deoxynucleotidyl transferase) ***(38)***. This translocation was thought to be linked with basophilia as inv(16) was associated with eosinophilia; it is now known that there is an association, but it is not nearly so specific and no basophilia is detected in many cases. The breakpoint on chromosome 9 is at 9q34.3, which is distal to the breakpoint in the Philadelphia translocation; it involves a different gene, CAN instead of

ABL. However, the cytological appearance of the 9q+ is similar (Fig. 1). The prognosis is generally poor.

5.3. Cryptic Abnormalities in AML

Overall, a clone is found in approx 60% of cases of AML by conventional cytogenetic study. The genetic abnormality in most of the remaining 30% of cases has still to be determined. In some cases, cryptic rearrangements of the genes involved in the commonly occurring translocations already described have been demonstrated ***(39)***. A published study ***(40)*** of a large series of patients found a high incidence of rearrangements of the *ETO/AML1* genes, indicating the presence of a t(8;21) rearrangement in the absence of any cytogenetic evidence of abnormality, or masked by the presence of a different abnormality. Similar results were found for cryptic inv(16)(p13q22) ***(41)***. However, several laboratories were unable to confirm these findings ***(42)*** and it now seems likely that the incidence of cryptic versions of these translocations is rare.

5.4. Secondary MDS and AML

It is a tribute to modern cancer treatments that increasing numbers of patients are cured or have a greatly extended survival. However, the downside is that a smaller but similarly increasing number of patients is living long enough to suffer unwanted side effects of that treatment. Whether or not some patients are inherently at greater risk of developing more than one kind of malignancy, there is an inescapable association between intensive, genotoxic therapy and the emergence of a second cancer. A patient's bone marrow is constantly active and the DNA of dividing bone marrow cells is susceptible to damage; consequently, MDS and AML are the most common secondary malignancies. These tend to fall into one of two classes, depending on the type of treatment for the primary disease:

1. Cases of MDS/AML that are secondary to exposure to alkylating agents, particularly when the exposure has been to both chemotherapy and radiotherapy. This typically arises at least 3 yr after

commencement of exposure, although this latent interval can be much shorter after very intensive treatment, such as for bone marrow transplant. Cytogenetically, abnormalities of chromosomes 5 and 7 are most common, usually as part of a complex clone. These patients usually have a very poor response to treatment.

2. AML secondary to treatment by epipodophyllotoxins. In this event, the time between exposure and diagnosis is often < 2 yr. Cytogenetically, abnormalities involving 11q23 are most frequent; however, also common are translocations involving 21q22, including t(8;21)(q22;q22), and also the t(15;17)(q24;q21) that is typical of AML M3. In all these patients the prognosis is considerably better, being very similar to that of primary AML.

6. Acute Biphenotypic Leukemia

Mention is made here of a newer grouping of AMLs, those that are shown by immunology to express unusually high levels of lymphocyte cell surface markers. This is termed biphenotypic AML, and it is usually associated with a relatively poor prognosis. However, this prognosis is more likely to be a consequence of the presence of poor-risk cytogenetic abnormalities than being directly related to the phenotype ***(43)***, as the most common cytogenetic abnormality is the Philadelphia translocation, t(9;22)(q34;q11) ***(44)***. The t(8;21)(q22;q22) is also included in some series of biphenotypic leukemias, largely because it is commonly associated with a lymphoid antigen, CD19.

7. Summary

Myeloid disorders do not usually present quite so many technical challenges to the cytogeneticist as does ALL: the chromosomes are often of a better quality, and white blood cell counts are not usually so high, except in CGL. Unlike in the chronic lymphoid disorders, there is no need for mitogens to include cell division. However, apart from the Ph in CGL, the overall frequency of detected clones is not so high. This has the consequence that a large proportion of patients is denied the diagnostic and prognostic benefit of knowing the cytogenetic abnormalities that are associated with their disease.

References

1. Nacheva, E., Holloway, T., Carter, N., Grace, C., White, N., and Green, A. R. (1995) Characterization of 20q deletions in patients with myeloproliferative disorders or myelodysplastic syndromes. *Cancer Genet. Cytogenet.* **80,** 87–94.
2. Third International Workshop on Chromosomes in Leukemia (1981) Report on essential thrombocythemia. *Cancer Genet. Cytogenet.* **4,** 138–142.
3. Aviram, A., Blickstein, D., Stark, P., et al. (1999) Significance of *BCR-ABL* transcripts in bone marrow aspirates of Philadelphia-negative essential thrombocythemia patients. *Leukemia Lymphoma* **33,** 77–82.
4. Singer, I. O., Sproul, A., Tait, R. C., Soutar, R., and Gibson, B. (1998) *BCR-ABL* transcripts detectable in all myeloproliferative states. *Blood* **92,** 427a.
5. Marasca, R., Luppi, M., Zucchini, P., Longo, G., Torelli, G., and Emilia, G. (1998) Might essential thrombocythemia carry Ph anomaly. *Blood* **91,** 3084.
6. Hackwell, S., Ross, F., and Cullis, J. O. (1999) Patients with essential thrombocythemia do not express *BCR-ABL* transcripts. *Blood* **93,** 2420–2421.
7. Cervantes, F., Rozman, M., Urbano-Ispizua, A., Monserrat, E., and Rozman, C. (1990) A study of prognostic factors in blast crisis of Philadelphia chromosome positive chronic myeloid leukemia. *Br. J. Haematol.* **76,** 27–32.
8. Huntly, B. J., Reid, A. G., Bench, A. J., et al. (2001) Deletions of the derivative chromosome 9 occur at the time of the Philadelphia translocation and provide a powerful and independent prognostic indicator in chronic myeloid leukemia. *Blood* **98,** 1732–1738.
9. Heim, S. and Mitelman, F. (1995) *Cancer Cytogenetics,* 2nd edit. Pub. A. R. Liss, New York, p. 38.
10. Fioretos, T., Strombeck, B., Sandberg, T., et al. (1999) Isochromosome 17q in blast crisis of chronic myeloid leukemia and in other hematologic malignancies is the result of clustered breakpoints in

17p11 and is not associated with coding TP53 mutations. *Blood* **94,** 225–232.

11. Bennett, J. M., Catovsky, D., Daniel, M. T., et al. (1982) The FAB Co-operative Group. Proposals for the classification of the myelodysplastic syndromes. *Br. J. Haematol.* **51,** 189–199.
12. Van den Berghe, H., Vermaelen, K., Mecucci, C., Barbieri, D., and Tricot, G. (1985) The 5q- anomaly. *Cancer Genet. Cytogenet.* **17,** 189–255.
13. Passmore, S. J., Hann, I. M., Stiller, C. A., et al. (1995) Pediatric myelodysplasia: a study of 68 children and a new prognostic scoring system. *Blood* **85,** 1742–1750.
14. Reiter, A., Hehlmann, R., Goldman, J. M., and Cross, N. C. P. (1999) The 8p11 myeloproliferative syndrome. *Medizin. Klin.* **94,** 207–210.
15. Appelbaum, F. R., Barrall, J., Storb, R., et al. (1987) Clonal cytogenetic abnormalities in patients with otherwise typical aplastic anemia. *Exp. Hematol.* **15,** 1134–1139.
16. Moormeier, J. A., Rubin, C. M., Le Beau, M. M., Vardiman, J. W., Larson, R. A., and Winter, J. N. (1991) Trisomy 6: a recurring cytogenetic abnormality associated with marrow hypoplasia. *Blood* **77,** 1397–1398.
17. Bennett, J. M., Catovsky, D., Daniel, M. T., et al. (1976) Proposals for the classification of acute leukemias (FAB co-operative group). *Br. J. Haematol.* **33,** 451–458.
18. Bennett, J. M., Catovsky, D., Daniel, M. T., et al. (1985) Proposed revised criteria for the classification of acute myeloid leukemia. *Ann. Intern. Med.* **103,** 626–629.
19. Mehta, A. B., Bain, B. J., Fitchett, M., Shah, S., and Secker-Walker, L. M. (1998) Trisomy 13 and myeloid malignancy—characteristic blast cell morphology: a United Kingdom cancer cytogenetics group survey. *Br. J. Haematol.* **101,** 749–752.
20. Langabeer, S. E., Grimwade, D., Walker, H., et al. (1998) A study to determine whether trisomy 8, deleted 9q and trisomy 22 are markers of cryptic rearrangements of PML/RARalpha, AML1/ETO and CBFB/MYH11 respectively in acute myeloid leukemia. *Br. J. Haematol.* **101,** 338–340.
21. Rege, K., Swansbury, G. J., Atra, A. A., et al. (2001) Disease features in acute myeloid leukemia with t(8;21)(q22;q22). Influence of age, secondary karyotype abnormalities, CD19 status, and extramedullary leukemia on survival. *Leukemia Lymphoma* **40,** 67–77.

22. Stock, A. D., Dennis, T. R., and Spallone, P. A. (2000) Precise localization by microdissection/reverse ISH and FISH of the t(15;17)(q24;q21.1) chromosomal breakpoints associated with acute promyelocytic leukemia. *Cancer Genet. Cytogenet.* **119,** 15–17.
23. Grimwade, D., Walker, H., Oliver, F., et al. (1998) The importance of diagnostic cytogenetics on outcome in AML: analysis of 1,612 patients entered into the MRC AML 10 trial. *Blood* **92,** 2322–2333.
24. Nucifora, G., Larson, R. A., and Rowley, J. D. (1993) Persistence of the 8;21 translocation in patients with acute myeloid leukemia type M2 in long-term remission. *Blood* **82,** 712–715.
25. Hiorns, L. R., Swansbury, G. J., Mehta, J., et al. (1997) Additional abnormalities confer worse prognosis in acute promyelocytic leukemia. *Br. J. Haematol.* **96,** 314–321.
26. Pantic, M., Novak, A., Marisavljevic, D., et al. (2000) Additional chromosome aberrations in acute promyelocytic leukemia: characteristics and prognostic influence. *Medical Oncology* **17,** 307–313.
27. Tobal, K., Saunders, M. J., Grey, M. R., and Yin, J. A. (1995) Persistence of RAR alpha-PML fusion mRNA detected by reverse transcriptase polymerase chain reaction in patients in long-term remission of acute promyelocytic leukemia. *Br. J. Haematol.* **90(3),** 615–618.
28. Licht, J. D., Chomienne, C., Goy, A., et al. (1995) Clinical and molecular characterization of a rare syndrome of acute promyelocytic leukemia associated with translocation (11;17). *Blood* **85,** 1083–1094.
29. Wong, K. F. and Kwong, Y. L. (1999) Trisomy 22 in acute myeloid leukemia: a marker for myeloid leukemia with monocytic features and cytogenetically cryptic inversion 16. *Cancer Genet. Cytogenet.* **109,** 131–133.
30. Betts, D. R., Rohatiner, A. Z. S., Evans, M. L., et al. (1992) Abnormalities of chromosome 16q in myeloid malignancy: 14 new cases and a review of the literature. *Leukemia* **6,** 1250–1256.
31. Hanslip, J. I., Swansbury, G. J., Pinkerton, R., and Catovsky, D. (1992) The translocation t(8;16)(p11;p13) defines an AML subtype with distinct cytology and clinical features. *Leukem. Lymphoma* **6,** 479–486.
32. Sorour, A., Brito-Babapulle, V., Smedley, D., Yuille, M., and Catovsky, D. (2000) Unusual breakpoint distribution of 8p abnormalities in T-prolymphocytic leukemia: a study with YACS mapping to 8p11–p12. *Cancer Genet. Cytogenet.* **121,** 128–132.
33. Abdou, S. M. H., Jadayel, D. M., Min, T., et al. (2002) Incidence of *MLL* Rearrangement in acute myeloid leukemia, and *CALM-AF10* fusion in AML M4. *Leukemia & Lymphoma* **43,** 89–95.

34. Sait, S. N., Brecher, M. L., Green, D. M., and Sandberg, A. A. (1988) Translocation t(1;22) in congenital acute megakaryocytic leukemia. *Cancer Genet. Cytogenet.* **34,** 277–280.
35. Carroll, A., Civin, C., Schneider, N., et al. (1991) The t(1;22) (p13;q13) is nonrandom and restricted to infants with acute megakaryoblastic leukemia: a Pediatric Oncology Group Study. *Blood* **78,** 748–752.
36. Schoch, C., Haase, D., Fonatsch, C., et al. (1997) The significance of trisomy 8 in de novo acute myeloid leukemia: the accompanying chromosome aberrations determine the prognosis. *Br. J. Haematol.* **99,** 605–611.
37. Secker-Walker, L. M., Mehta, A., Bain, B., and Martineau, M. (1995) Abnormalities of 3q21 and 3q26 in myeloid malignancy: a United Kingdom cancer cytogenetic group study. *Br. J. Haematol.* **91,** 490–501.
37. Cuneo, A., Kerim, S., Vandenberghe, E., et al. (1989) Translocation t(6;9) occurring in acute myelofibrosis, myelodysplastic syndrome, and acute nonlymphocytic leukemia suggests multipotent stem cell involvement. *Cancer Genet. Cytogenet.* **42,** 209–219.
39. Hiorns, L. R., Min, T., Swansbury, G. J., Zelent, A., Dyer, M. J. S., and Catovsky, D. (1994) Interstitial insertion of retinoic receptor-α gene in acute promyelocytic leukemia with normal chromosomes 15 and 17. *Blood* **83,** 2946–2951.
40. Langabeer, S. E., Walker, H., Rogers, J. R., et al. (1997a) Incidence of AML1/ETO fusion transcripts in patients entered into the MRC AML trials. *Br. J. Haematol.* **99,** 925–928.
41. Langabeer, S. E., Walker, H., Gale, R. E., et al. (1997b) Frequency of CBFbeta/MYH11 fusion transcripts in patients entered into the U. K. MRC AML trials. *Br. J. Haematol.* **96,** 738–739.
42. Rowe, D., Cotterill, S. J., Ross, F. M., et al. (2000) Cytogenetically cryptic AML1/ETO and CBFB/MYH11 gene rearrangements: incidence in 412 cases of acute myeloid leukemia. *Br. J. Haematol.* **111,** 1051–1056.
43. Killick, S., Matutes, E., Powles, R. L., et al. (1999) Outcome of biphenotypic acute leukemia. *Hematologica* **84,** 699–706.
44. Carbonell, F., Swansbury, J., Min, T., et al. (1996) Cytogenetic findings in acute biphenotypic leukemia. *Leukemia* **10,** 1283–1287.

4

Cytogenetic Techniques for Myeloid Disorders

John Swansbury

1. Introduction

Chromosomes are prepared from dividing cells (mitoses), as at metaphase, just before division, they shorten and become recognizable, discrete units. The cells are arrested and accumulated in metaphase or prometaphase by destroying (e.g., with colcemid) the mitotic spindle which would separate the chromatids. The cells are treated with a hypotonic solution to encourage spreading of the chromosomes. They are then fixed, after which they can be stored indefinitely. Fixed cells are spread on slides and air-dried. They can be stained immediately, but are usually first treated to induce banding patterns on the chromosomes to assist in their identification.

2. Materials

Many of the reagents and chemicals used are harmful and should be handled with due care and attention. Always refer to the information provided by the manufacturer.

From: *Methods in Molecular Biology, vol. 220: Cancer Cytogenetics: Methods and Protocols*
Edited by: John Swansbury © Humana Press Inc., Totowa, NJ

Most of the solutions should be kept in the dark at about 4°C. The dilutions given here of most of the reagents are such that 0.1 mL may be conveniently added to a 10-mL culture.

Except for phytohemagglutinin (PHA), the solutions should be filter sterilized (using, e.g., a 0.22-µm Millipore filter).

1. Containers: Sterile, capped, plastic, 10-mL centrifuge tubes. The caps should be well fitting enough to prevent leakage of fixative. The author's laboratory uses Nunc Leyton tubes, which have particularly good caps that plug inside the tube as well as a screw fitting outside. These tubes are used both for the cultures and for the harvesting. However, using larger tubes (such as 20-mL Universal tubes) or tissue culture flasks also works well for cultures, and has the advantage of allowing a greater surface area at the interface between medium and cell pellet, which bone marrow cells seem to prefer.
2. Pipets: Plastic, disposable. For setting up cultures, the pipets must be sterile; for harvesting cultures they do not need to be sterile. Glass pipets should not be used because of the risk of needlestick injury and also because fixed cells will adhere to the glass.
3. Medium: RPMI 1640 (GIBCO) with Glutamax is recommended. Many other media may be used successfully, such as McCoy's 5A and Ham's F10, but RPMI was developed specifically for leukemic cells. The medium commonly used for PHA-stimulated cultures of blood lymphocytes, TC199, is less suitable for bone marrow cultures. To each 100-mL bottle add antibiotics (e.g., 1 mL of penicillin + streptomycin) and 1 mL of preservative-free heparin. If the medium does not contain Glutamax then L-glutamine should be added (final concentration 0.15 mg/mL); this is an essential amino acid that is unstable and has a short life at room temperature.
4. Serum: Fetal calf serum; the proportion routinely added is 15 mL of serum to 100 mL of medium.
5. Blocking agent: *5-Fluoro-2-deoxyuridine (FdUr): stock solution: 0.25 mg FdUr and 96 mg uridine made up to 100 mL with distilled water, giving final concentrations of 0.1 µ*M* and 4.0 µ*M*. Store frozen in 2-mL volumes; once thawed the effectiveness declines after a week.
6. Releasing agent: thymidine; 10 µ*M* stock solution: 0.05 g in 100 mL of distilled water. Store at –20°C in 2-mL aliquots. The thawed solution keeps at 4°C for about 1 mo; do not refreeze.

7. Arresting agents: colcemid (also called demecolchicine, from deacetylmethylcolchicine). Stock solution 1 μg/mL, or as provided by the supplier. Its effect is quantitative; that is, the more cells present in the culture, the more colcemid will be needed. The amount recommended here should be adequate for the standard culture containing 10^7 cells. Arresting agents act by preventing spindle formation and so the chromosomes remain dispersed in the cytoplasm. Another effective and widely used arresting agent is colchicine; this has an irreversible effect, whereas colcemid can be washed out of the culture if necessary.
8. Hypotonic solution: 0.075 *M* potassium chloride (KCl). Use 5.59 g of KCl and make up to 1 litre of aqueous solution. Use at 37°C. Note that the effectiveness does not derive from just the osmolarity: the K^+ ions have a physiological action, so no advantage is obtained by diluting further. With longer chromosomes, twisting or overlapping can be a problem and the use of 19 parts of KCl to 1 part of 0.8% sodium citrate is sometimes helpful.
9. Fixative: Three parts absolute methanol and one part glacial acetic acid. This should be freshly prepared just before use although it may be kept for a few hours if chilled.
10. 2.5% Trypsin: Stored frozen in 1-mL aliquots. Diluted 1:50 in buffer (Ca^{2+}- and Mg^{2+}-free, e.g., Hank's buffered salt solution) when required.
11. Phosphate-buffered saline (PBS): pH 6.8, used for diluting stain.
12. PHA (*see* Chapter 9).
13. Slides: The frosted-end variety is preferable for convenience of labeling. The slides must be free of dust and grease. Specially cleaned slides may be bought, otherwise wash in detergent, rinse well in water, then in dilute hydrochloric acid and alcohol.
14. Stains: Wright's stain. (Giemsa and Leishman's stains are also suitable.) This is usually obtained as a powder. Cover a flask with aluminum foil, and insert a magnetic stirrer. Add 0.5 g of stain and 200 mL of methanol. Stir for 30 min. Filter through filter paper into a foil-coated bottle. Close the lid tightly, and store the bottle in a dark cupboard for at least a week before use. The stain is diluted immediately before use 1:4 with pH 6.8 buffer.
15. Coverslips: 22 × 50 mm, grade 0 is preferred but thickness up to grade 1.5 is usually satisfactory.
16. Mounting medium: Gurr's neutral mounting medium is routinely used in this laboratory; in our experience it does not leach stain if it

not diluted with xylene. Other suitable mountants are XAM, DPX, Histamount, and so forth. Mounting slides has the advantage of protecting delicate chromosome spreads from dust and scratches. However, if it likely that a slide might need to be destained and processed for analysis by fluorescence *in situ* hybridization (FISH), then it should not be mounted.

17. Incubator: Ideally, a CO_2-controlled, humidified incubator at 37°C should be used, although good results are usually obtained with a simple incubator that has only temperature control. Alternatively, cultures can be gassed with 4% CO_2 in air (which also helps to maintain the pH of the culture medium if it is bicarbonate buffered). An increased partial pressure of CO_2 is not as important for cell growth as a decreased partial pressure of oxygen: 2.5% oxygen may be optimal for longer cultures.
18. Centrifuge: This should have buckets that can be sealed, to prevent the dispersal of aerosol into the laboratory by the air that blows through the centrifuge to keep the motor cool. It is safest to carry the closed bucket to a laminar flow cabinet before opening it to remove the tubes.

    The appropriate speed and duration of centrifugation depends on the type of centrifuge that the laboratory uses; the longer the rotor arm, the greater is the centripetal force, so the speed or time can be reduced. It is important to ensure that sufficient time is given for the white cells to be collected, as they generally take longer than the red cells. This is particularly so after the hypotonic step, described in **Subheading 3.4.**, step 5, so setting the tubes to centrifuge for a few minutes longer after this step can be worthwhile. In the author's laboratory, centrifugation takes 11 min at 12,000 rpm; the most effective speed and time would need to be determined for other centrifuges.
19. Laminar flow cabinet: A cabinet with vertical airflow is most suitable, as this serves both to protect the sample from contamination and to protect the cytogeneticist from infection.

3. Methods

3.1. Receiving and Assessing the Sample

When a sample is received in the laboratory it should be checked to ensure (1) that the patient's ID on the container matches that on the referral form; (2) that it is the appropriate sample for the investigation required; (3) that it is of an adequate quantity and quality, not leaking or clotted. The sample details will be recorded on computer or in a laboratory daybook, and a decision made about the type of cultures that will be used, depending on the diagnosis and on the type of investigation wanted.

If the sample arrives without medium then add warmed medium immediately to make up to about 10 mL. If the sample arrived in medium then it is better to centrifuge the sample and resuspend it in fresh medium and serum. Centrifuging the sample at this stage also has the advantage of revealing approximately how many white cells are present. The white cell-rich layer should be removed with a sterile pipet if there is a large proportion of red cells, as they will interfere with some of the later harvesting and processing steps. However, if the white cell count is very low, it is better to use the entire sample rather than risk losing some cells by attempting to separate them.

If the laboratory has access to a cell counter, then it is worth using it to identify those cases with particularly high or low cell counts. It is easy to over-inoculate cultures from patients with chronic myeloid leukemia (CML) and occasionally other types of leukemia if the white cell count is very high; the final dilution should be about $1–2 \times 10^6$/mL. Adding too many cells usually results in failure to obtain any useful divisions at all. Under-inoculation is less serious, but if the cell count is very low then use smaller tubes and set up 5-mL or 2-mL cultures.

If the sample has clotted, *see* **Note 1**.

3.2. Choice of Cultures

It has been shown that in samples of myeloid disorders, erythropoietic divisions predominate in the first few hours of culture, with granulopoietic divisions appearing subsequently (*see* Chapter 2). This corresponds to the observation that for the rare erythroleukemia (AML-M6), short-term cultures (harvested the same day that the sample was taken) are more likely to have clonal divisions. For all other AMLs the sample should be cultured for at least 16 h (i.e., overnight; *see* **Note 2**).

As described in Chapter 2, several cultures should be set up if there is sufficient material is available, because the cell cycle time is unpredictably affected by the disease. For this description of technique, it will be assumed that the supplied sample had enough cells for four cultures: 24-h, overnight colcemid, FdUr-blocked ***(1)***, and a PHA-stimulated culture to check the constitutional karyotype in case it becomes necessary to do so. In the author's laboratory, it is also customary to set up a further culture, if there is sufficient material, blocked with excess thymidine as described in Chapter 9, and to have two cultures blocked with FdUr that are released at different times. If the sample is very small, consider using half-volume cultures, or else reduce the number of cultures set up. In most cases the order of priority for cultures is the 24-h culture, then the colcemid overnight, then the FdUr-blocked culture. The PHA culture is least important, as it is usually possible to obtain some blood from the patient at a later stage if it is needed. The PHA-stimulated culture is described in more detail in Chapter 9.

3.3. Setting Up Cultures

All handling of unfixed tissues should take place in a laminar flow cabinet, and proper protective clothing should be worn, gloves and laboratory coat being the minimum.

1. When the sample is received, add some warmed culture medium and leave the sample in the incubator until a convenient time for setting up the cultures. The FdUr-blocked culture in particular benefits if the cells have had a few hours growth before being blocked. Perform a cell count, enter the sample in the laboratory record system, and print out labels for the cultures.
2. Fix the labels to the culture tubes, add the appropriate measured amount of cell suspension such that the cell count does not exceed 2×10^7 per tube, and then add more medium to bring the volume to nearly 10 mL.
3. The PHA can be added at any time, but wait until the end of the afternoon before adding colcemid to one culture and FdUr to another. Mix gently but thoroughly, and then place the tubes in the incubator.

It can help to stand them at an angle, rather than upright, as this increases the surface area of the deposit and helps to reduce local exhaustion of the medium.

4. Next morning, harvest the tube that had overnight colcemid, add thymidine to release the tube that was blocked with FdUr (*see* **Note 3**), and add colcemid to the 24-h-culture tube.
5. Harvest the 24-h culture after 1 or 2 h, and harvest the blocked culture after 4 h (*see* **Note 4**).

This procedure is presented as a flow diagram in **Table 1**.

3.4. Harvesting

Samples must be in capped tubes during centrifugation and the centrifuge buckets should have secure lids to avoid aerosol dispersion. As with setting up, sample processing should be done in a laminar flow cabinet.

Note: A few minutes of extra time spent on careful harvesting to get the best possible quality chromosomes can save hours later when it comes to the analysis! Once cells have been fixed, there is little that can be done to improve their quality.

1. Place the hypotonic KCl solution in an incubator to warm up.
2. Centrifuge the tubes to collect all the cells. Ensure that the tubes are placed symmetrically in the centrifuge buckets, so that they are balanced. Use extra, dummy tubes if necessary.
3. Remove the supernatant culture medium, using a plastic pipet (clean but not necessarily sterile at this stage), and being careful not to disturb the cell pellet. Put the supernatant into a waste container for safe disposal later.
4. Add up to 10 mL of warmed hypotonic KCl. Replace the cap of the tube and mix gently but thoroughly by inversion. Leave in the incubator for 15 min.
5. Centrifuge again. This may need 1 or 2 min more than previously, as the swollen cells take longer to move to the bottom of the tube.
6. Remove the supernatant. Add two or three drops of hypotonic KCl and thoroughly resuspend the cell pellet by tapping the tube. If using the vortex mixer, set it at a low speed for this stage, as fierce mixing will lead to rupture of cell membranes and loss of chromosomes.
7. Make up the fixative, 3:1 methanol–glacial acetic acid.

Table 1
Scheme for Setting Up Cultures

1	Label four sterile culture tubes with the date, the case ID, and each culture type:			
2	24-h	CON (colcemid overnight)	FdUr	PHA
3	Add a measured volume of sample, with the appropriate number of white cells.			
4	Add the culture medium to make up to 10 mL.			
5	Leave in the incubator until the end of the afternoon.			
6		Add 100 µL of colcemid.	Add 100 µL of FdUr	Add 100 µL of PHA
7	Return the tubes to the incubator, and stand them at an angle rather than upright.			
8	Next morning, add 100 µL of colcemid. Mix the contents by inverting the tube.	Next morning, start harvesting	Next morning, add 100 µL of thymidine and 100 µL of colcemid (*see* **Note 1**). Mix the contents by inverting the tube.	Three days later, add 100 µL colcemid (*see* **Note 2**). Mix the contents by inverting the tube.
9	Return to the incubator for 1 h, then start harvesting.		Return to the incubator for 4 h, then start harvesting (*see* **Notes 2** and **3**).	Return to the incubator for 1 h, then start harvesting.

8. Add a few drops of fixative to the cells and mix well. The first few drops of fix are the most important and can have a large effect on chromosome quality.
9. Add further small quantities of fixative, mixing after each, until about 2 mL have been added.
10. Add fixative up to 10 mL. Replace the cap of the tube and mix vigorously by inversion.

11. Centrifuge the tubes, then assess the cell deposit. If the volume of red blood cells was large and inadequately dispersed in hypotonic solution, they may fix into a semitransparent or gelatinous mass. This usually interferes with spreading and banding, so remove the fixative and resuspend the cells in hypotonic KCl for a few minutes, then centrifuge and repeat the fixation.
12. Remove the fixative. At this stage and afterwards, it is usually possible to remove the supernatant by gently pouring it out of the tube rather than pipetting. The cell deposit is usually very small and should stay safely at the bottom of the tube. If there is a large cell deposit, it indicates that the culture was over-inoculated with too many cells; this usually results in there being few divisions. If a large deposit is found in the first culture harvested, consider splitting the other cultures and diluting the cells by adding more medium.
13. Add fresh fixative and mark a spot on the cap of the tube.
14. Repeat the centrifugation and the change of fixative at least twice more, until the solution is colorless and the cell deposit is white. Mark the cap each time to keep record of the number of fix changes.
15. If there was a large amount of fat in the specimen, it may cause problems with the spreading. It can be removed by washing once with Carnoy's fixative (ethanol 60%, chloroform 30%, acetic acid 10%).
16. Stored at –20°C, the DNA of the chromosomes will remain good for several years for chromosome spreads or for molecular analysis. If using archived material to make fresh slides, give the cells two changes of fresh fix before spreading. Note that ordinary, thin plastic test tubes are not suitable for long-term storage of cells in fixative: the plastic degrades and eventually disintegrates. It is better to use the smaller 1.8-mL polypropylene tubes (e.g., Nunc™) that are specially designed for cryopreservation.

3.5. Spreading

Atmospheric factors affect spreading, and it has been shown that the air temperature should be about 25°C and the humidity should be about 50–55% ***(2)***.

1. Sometimes a sample that was not clotted becomes clotted during processing. There is nothing that can be done once the cells are fixed. Remove any lumps or strands of material before spreading the remaining cells.

2. Change the fixative shortly before spreading, then centrifuge again and add a few drops of fresh fixative to obtain a slightly cloudy suspension. Judging the correct dilution comes with experience: if the cells are not sufficiently dispersed, then the chromosomes will not spread or stain properly; if the dilution is too great then time will be wasted in screening nearly blank slides.
3. Generally just two slides are spread for each culture. Each slide should be labeled with the date, the patient's ID, the type of culture, and a serial number. There are many ways of spreading the cell suspension on slides; three are described below. The author's preference is to use wet slides.
   a. Using cold-frosted slides, kept in a freezer: Take a pair of slides from the freezer. Add two well-spaced drops of cell suspension to each slide. Wave the slides briefly in the warm moist air above a Bunsen burner flame adjusted to be clear but not fierce. Label the slides immediately.
   b. Using dry slides: Label the dry slides and place flat on a slide tray. Add two drops of cell suspension, then immediately blow gently, or "huff," on the slides.
   c. Using wet slides: Label the dry slides. Dip slides singly or in pairs into a beaker of distilled water; ensure that there is an even film of water on each slide. Add two drops of cell suspension to each slide, then gently flick off the excess water. Stand upright to let the remaining water drain away.
4. Leave the slides to dry at room temperature. It is important that the fixative does not dry too quickly, as this will not give the chromosomes a chance to spread out. Some cytogeneticists like to add a drop of fresh fixative immediately after spreading, to slow down the drying out time.
5. Check the first slides using phase-contrast microscopy, if possible, for cell density and chromosome spreading. If these are not optimal, try adjusting the dilution, or varying the spreading procedure (*see* **Note 5**).

3.6. Banding

There are several methods of producing bands on chromosomes. Band patterns are broadly grouped into G (Giemsa) bands and R (reverse) bands which are largely complementary *(3)*. G-banding is the most widely used. R-banding is the preferred type in some parts of Europe; it has the advantage of showing up some abnormalities more clearly, but the disadvantages of poor chromosome morphology, the need for a fluorescence microscope, and rapid fading.

Other banding methods are available that are specific for identifying certain limited chromosome regions, such as C-banding and G11 banding; however, these have become largely redundant since the introduction of FISH.

Banding is not very effective until the chromosome spreads have "aged" for a few days. In practice, few laboratories can afford to wait this long (*see* **Note 6**), so one of the following procedures may be used:

a. Incubate the slides in an oven at 60°C overnight.
b. Incubate the slides in an oven at 90°C for 1 h.
c. Incubate the slides in an oven at 100°C for 20 min.
d. Microwave the slides for 5 min.
e. Stain the slides, dry, destain in fixative, then allow to dry.

For options a–d, allow the slides to cool for a few minutes before starting the banding; do not delay too long. If the slides are not to be banded at once, store them at room temperature, protected from dust; if they are needed later, put them back in the oven for at least 10–15 min before they are banded.

1. Set up a series of Coplin jars (*see* **Notes 7** and **8**):
   a. Buffer, pH 6.8.
   b. Trypsin, 0.5 mL in 50 mL of Hank's buffered salts solution.
   c. Dilute serum (about 2%) in buffer.
   d. Buffer.

2. Examine the slides under phase-contrast microscopy, if possible, and choose one with plenty of divisions to serve as a test slide.
3. Rinse the slide (or two slides back to back) briefly in buffer.
4. Dip about one third of the slide into the trypsin solution, gently waving the slide in the solution. After 10 s lower the slide so a further one third is immersed, and after another 10 s the last third, leaving the whole slide immersed for 10 s. Once the appropriate time in trypsin has been determined for that day, this stepwise immersion

may not be necessary.

5. Rinse the slide for a few seconds in the dilute serum to arrest the action of the trypsin, then rinse in the buffer.
6. Remove the slide, drain off the buffer, then place the slide flat, face upwards, on a pair of glass rods across a sink or large dish. (However, slides can be stained vertically, in a Coplin jar, if necessary.)
7. Prepare the stain, approx 2.5 mL per slide, in the proportion of 1 part of Wright's stain stock solution to 4 parts of pH 6.8 buffer. Mix well with a pipet, then place on the slide.
8. After 5 min, tip the slide to pour off the stain, rinse it briefly in tap water, then place it in buffer to differentiate the bands by washing out some of the stain.
9. After 2 min, remove the slide and blot it dry, face down, on a clean paper towel. Blotting slides may seem risky, as the chromosomes are easily scratched, but so long as the slide is lifted straight off the towel (without slipping), there should not be any scratches. It is important not to use paper towels or tissues that leave a shower of fragments stuck to the slide.

   The alternative is to flick as much buffer as possible off the slides, then stand them to drain dry. However, this usually results in stain being leached out of patches of the slide where water droplets remained.
10. Use an electric fan to help the slide to dry thoroughly.
11. When the slide is completely dry, place two or three drops of mounting medium on a coverslip and then place it face down on the slide. Examine the test slide under high power on the microscope and assess the quality of the banding in each third (*see* **Note 9**) to determine the best length of time for immersion in trypsin. At the same time, assess the quality of the stain and adjust the concentration, the exposure time, and the differentiation time if necessary. The slide can be de-stained and re-stained as described in **Note 10**.

    When a satisfactory result has been obtained, use the same procedure for all the other slides. If possible, check a slide from each case before staining other slides, as the banding will vary according to the length of the chromosomes.
12. Leave the slides to dry overnight if possible before screening and analyzing the metaphases. Immersion oil and soft mounting medium tend to mix and produce a sticky compound that adheres to the microscope objective lens and makes it impossible to focus properly.

13. When the microscopic analysis of each slide has been finished, place the slide face down on tissue to remove the immersion oil, as otherwise it may seep under the coverslip and leach the stain. Do not leave the slides exposed to direct sunlight; stored in the dark they should keep in good condition for several years.

4. Notes

1. Dealing with clotted samples. These need some intervention to release the white cells that will be trapped in the fibrin, as analyzable metaphases cannot be obtained from a fixed lump. If the clot is soft, it can be broken down by gentle agitation with a sterile pipet. Remove the supernatant and use it to set up some cultures. Suspend the remaining fragments in a trypsin solution (1 mL of trypsin in 100 mL of Hanks' buffered salt solution or in RPMI culture medium without any added serum) and leave it for 30 min at 37°C in the incubator. Assess the effect of this digestion by gently agitating with a sterile pipet again; if it has been successful, centrifuge the suspension, remove the trypsin solution, and resuspend the cells in complete medium.

   If the clot has become hard, try a longer exposure to trypsin, although even this is likely to have limited success; it may be better to contact the clinician to see if some other sample may be available.

   If the hard clots are in the form of long strings or fragments, they were probably formed inside the syringe when the sample was being taken. This problem can usually be prevented by preloading the syringe with heparin before aspirating the sample. If there is one large clot, then it probably happened after the sample was put into a collection tube, and is more common with acute lymphoblastic leukemia (ALL) and AML M3 (promyelocytic leukemia) than with other kinds of AML.
2. The necessity for overnight cultures in AML can be inconvenient, as samples may be received just before a holiday or a weekend. In the author's laboratory, it is the policy that as far as possible samples are harvested at the normal times, even if it means staff coming in to work on a day when the laboratory is usually closed. An alternative is to set up the cultures and then place them in a refrigerator at 4°C; on the day before the laboratory reopens, a member of staff calls in to move the cultures from the refrigerator to the incubator.

3. Previously published protocols for blocked cultures recommend adding the releasing agent next morning, then leaving the cultures for some hours before adding colcemid, with harvesting starting a short time (10–20 min) later. This works well for normal, PHA-stimulated lymphocytes, when the optimum time for harvesting can be predicted. However, in the author's laboratory it has been found that adding the colcemid with the releasing agent can still sometimes produce long chromosomes, and has the advantage that if the wave of cell divisions happened to peak earlier then at least these divisions will be collected even if the chromosomes are short.
4. Reported protocols recommend that release of the blocking agent should be followed by approx 7–8 h of incubation before adding colcemid for 10–15 min before harvesting. In practice, it seems that most laboratories use only 4 h between release and addition of colcemid.
5. An occasionally useful variation is to spread from 60% aqueous acetic acid, which usually gives fair spreading of chromosomes but possibly at the expense of reduced banding quality. Another variation worth trying if the chromosomes will not separate is spreading from fixative made in a ratio of 2:1 instead of 3:1.
6. In the author's laboratory, the routine is to spread the slides on the day that the culture was harvested, and incubate one slide (or both) overnight in an oven at 60°C. Next morning, allow the slides to cool while the trypsin solution is being warmed. The slides are banded and stained, then mounted. They are then ready for immediate assessment. If the laboratory is already busy with other work and cannot analyze these slides immediately, then at least they are ready in case the clinician telephones to ask for an urgent result.
7. Some laboratories include an initial step of placing the slide in hydrogen peroxide diluted 1:3 with tap water, for 1 min; this is said to remove some of the cell cytoplasm and debris and give clearer banding.
8. Some laboratories prefer to use ice-cold solutions rather than have them at room temperature.
9. The time needed in trypsin immersion is very variable and may need to be adjusted for each case. It is affected by variations in spreading technique, age of the slides, degree of contraction of the chromosomes, general chromosome morphology, and so forth. If the chromosomes have been in trypsin too long, they will be too swollen; if

not long enough, then there will be poor banding with little contrast. If the even the bottom third of the slide, which had the longest time in trypsin, has not banded satisfactorily, then repeat the whole process with another test slide, but using longer times in the trypsin solution. For some slides it is not unknown to need up to a minute. A slide can be destained and taken through trypsin again if necessary; *see* **Note 10**.

It is sometimes possible to check the length of time needed by watching a metaphase under phase contrast while the trypsin is acting; as soon as the chromatids start to swell the trypsin action is halted.

10. It is occasionally necessary to remove the coverslip from a slide to destain and reband the chromosomes. If the mounting medium is still soft enough, gently and carefully slide off the coverslip. Stand the slide upright in a jar of fixative. Depending on how soft the mounting medium is, it needs to be left for a time ranging from a few hours to several days. If it was not removed previously, the coverslip will eventually become loose and fall off the slide. The mounting medium will turn white and it can be gently peeled off the slide, leaving the cells in place. Rinse the slide in fresh fixative, allow to dry, and then rinse in buffer before exposing to further trypsin and/or restaining.

References

1. Webber, L.M. and Garson, O.M. (1983) Fluorodeoxyuridine synchronisation of bone marrow cultures. *Cancer Genet. Cytogenet.* **8,** 123–132.
2. Wheater, R.F. and Roberts, S.H. (1987) An improved lymphocyte culture technique: deoxycytidine release of a thymidine block and use of a constant humidity chamber for slide making. *J. Med. Genet.* **24,** 113–115
3. ISCN (1995) *An International System for Human Cytogenetic Nomenclature.* (Mitelman, F., ed.), Karger, Basel, 1995.

5

Acute Lymphoblastic Leukemia

Background

John Swansbury

Unlike the situation in acute myeloid leukemia, in which there are at least eight morphological French–American–British (FAB) subtypes identified by the predominant cell cytology, in acute lymphoblastic leukemia (ALL), there are only three *(1)*. Furthermore, for most practical purposes, FAB types L1 and L2 are almost indistinguishable. Of more importance than the morphology of the cells in ALL is the immunophenotype. Apart from identifying an entirely separate T-cell type of ALL, immunology is able to identify cells at various stages of B-cell lineage maturation, including pro-B, pre-pre-B, pre-B, common (cALL), and mature B stages. The last stage constitutes FAB type L3, which has a very close affinity with Burkitt lymphoma; these two disorders share identical cytogenetic abnormalities and both are now formally termed Burkitt cell lymphoma (BL).

1. The Ploidy Classification

The chromosomes of the abnormal cells in samples from many patients with ALL have long had a notoriously poor morphology, especially those in a high hyperdiploid clone. Full analysis was (and

From: *Methods in Molecular Biology, vol. 220: Cancer Cytogenetics: Methods and Protocols*
Edited by: John Swansbury © Humana Press Inc., Totowa, NJ

sometimes still is) impossible, and the cytogeneticist had to be content with simply counting how many chromosomes there were. It happened that this proved to have some clinical value. Many of cases of childhood ALL were found to have cells with more than 50 chromosomes, and, more importantly, these children responded well to treatment, with high remission rates and many having a long disease-free survival *(2)*. These early observations led to the proposal of a "ploidy" classification that has been subsequently extended and still has clinical significance:

- *Diploid*: 46 normal chromosomes (22 pairs of autosomes plus two X chromosomes in females or an X and a Y in males).
- *Pseudodiploid*: 46 chromosomes with some abnormality. This term has become largely redundant now that more is known about structural abnormalities in ALL.
- *Near-haploid*: 25–29 chromosomes. There is always one of each chromosome, except for the Y. Near-haploid clones are rare but are associated with a very poor prognosis *(3)*. Sometimes a near-haploid clone undergoes a doubling of the entire chromosome set, resulting in a chromosome count in the high hyperdiploid range. However, the pattern of paired gains serves to distinguish it from the typical pattern of good-prognosis hyperdiploidy.
- *Low hypodiploid*: 30–40 chromosomes.
- *Hypodiploid*: 41–45 chromosomes.
- *Hyperdiploid*: 47–49 chromosomes. Some patients have a regular pattern, with gains of X, 6, 8, 10, 16, and/or 21. Others have no regular pattern of gains, and structural abnormalities are more often present.
- *High hyperdiploid*: 50–58 chromosomes, with a typical pattern of gains, the most common being X, 4, 6, 10, 14, 17, 18, 21 (often two and sometimes even three extra 21s). This is the most common type of abnormality found in childhood ALL, occurring in about 12% of reported cases. It has consistently been associated with a good prognosis *(4)*, even in the presence of less favorable structural abnormalities (*see* **Subheading 2.5.**). Some further prognostic subdivision of this group may be possible: cases with 54–58 chromosomes, especially with +4, +10, and +17, appear to have a favorable prognosis, and those with +5 a less good prognosis *(5)*.

It is not known how the common, specific pattern arises; the options appear to be (1) a sequential gain of the appropriate chromosomes; (2) a sequential loss of unwanted chromosomes from a triploid cell; or (3) a major nondisjunction event that happens to produce two daughter cells, one with all the right chromosomes gained, and the other cell with these chromosomes missing. Patients with high hyperdiploidy usually have other good prognostic factors: early pre-B-ALL type; SIg$^-$, CALLA+ and CD19+ immunophenotype; low white cell count. However, even those patients with high hyperdiploidy who do not have other favorable indicators tend to respond relatively well to treatment. Just as it is not yet known how typical high hyperdiploidy arises, it is also not known why patients with such clones respond well to treatment. It is a type of genetic abnormality that is not explained by current hypotheses about gene dosage effects. Patients with more than 50 chromosomes but without the typical pattern of gains do not have such a good prognosis.

- *Near-triploid*: 58–68 chromosomes. In ALL, clones in this group are not usually truly near-triploid (with three of every chromosome) but have the high hyperdiploid pattern plus further gains, with the most typical being 5 and 8 ***(6)***. The prognosis for this group is good.
- *Near-tetraploid:* About 92 chromosomes. These clones are rare, and tend to be associated with T-cell ALL *(7)*. Note that a few normal tetraploid divisions can be found in most bone marrow samples, and occasionally divisions can be seen with much higher ploidy levels, having hundreds of chromosomes. These cells usually derive from multinucleated megakaryocytes. Although some patients have unusually high numbers of tetraploid divisions, the rules for defining a clone must still be observed: there must be at least two divisions with some consistent numerical or structural abnormality. It is sometimes possible to trace an evolutionary path in the acquisition of abnormalities, with two copies of some, which therefore had occurred before the diploid set duplicated, and some being in single form and therefore having arisen afterwards.

2. Recurrent Structural Abnormalities

About 65% of cases of childhood ALL have a recognized "nonrandom" translocation, that is, one that has been found in several cases. A few are very common and have well described clinical associations. For others, publication of more data is still needed to identify their clinical associations. Abnormalities of chromosomes 3, 13, 15, 16, X, and Y are not common in childhood ALL (except for X often being part of the pattern of gains in high hyperdiploidy) and have not been associated with particular clinical features. The following subheadings describe some of the more characteristic abnormalities:

2.1. The t(1;19)(q23;p13.3)

A translocation t(1;19)(q23;p13.3) occurs in about 6% of all childhood cases of ALL, and in about 25% of cases with pre-B phenotype. There is a balanced form and a slightly more common unbalanced form in which there are two complete no. 1 chromosomes and the der(19)t(1;19) resulting in trisomy for 1q. Comparison of the two forms has detected no difference in presenting clinical characteristics except that the balanced form is associated with an older age group. The unbalanced form was found to have a better prognosis than the balanced form ***(8)***, but both types have been associated with a good prognosis when given modern intensive treatment ***(9)***. A breakpoint at 1q23 has also been observed in translocations with other chromosomes.

2.2. The t(4;11)(q21;q23)

A t(4;11)(q21;q23) occurs in all age groups but is the most common abnormality in infants with ALL. Some cases have been shown to be biphenotypic, having myeloid as well as lymphoid immunophenotype. Other cases have had other unusual features, being monocytoid, undifferentiated, T cell, or B cell. There is absence of the cALL antigen. This type of leukemia may therefore arise in a pluripotent stem cell. It tends to be associated with very high white cell counts, blasts in the peripheral blood, lymphadenopathy, hepatomegaly, and splenomegaly. The prognosis is very poor in infants, less so in children aged 2–10, and deteriorates again for patients over 40 years of age ***(10)***.

2.3. Deletion of 6q

Deletion of part of the long arms of a chromosome 6 occurs in a wide variety of lymphoid malignancies, including non-Hodgkin lymphoma (15% of cases studied). Various breakpoints have been described, including q13, q15, q21, and q23, all subject to observer error as the banding pattern on 6q has few landmarks when the morphology is suboptimal. It is not clear in most cases whether the deletion is interstitial or terminal. The commonly deleted segment has been claimed to be 6q21 ***(11)***.

In ALL, most cases with 6q- have been cALL or T-ALL; very few cases have been L3/BL. It occurs in about 10% of all cases of childhood ALL and has not been associated with any particular clinical characteristic. The prognosis in most studies has been intermediate.

2.4. Abnormalities of 8q24

The most significant abnormalities of chromosome 8 in ALL are three translocations involving 8q24 which have been found in almost all cases of L3/BL. Absence of one of these translocations has been established in a few well studied cases, and they do occasionally occur in other kinds of lymphoma, so the association with L3/BL is strong but not absolute. The gene at 8q24 is c-*MYC*. The three typical translocations are:

1. t(8;14)(q24;q32.3): This occurs in about 85% of cases of L3/BL. The breakpoint is close to the proximal end of c-*MYC*, and the partner gene on 14q32 is *IgH* (the immunoglobulin heavy gene).
2. t(8;22)(q24;q11): This occurs in about 10% of cases. The breakpoint is close to the distal end of c-*MYC*, and the partner gene on 22q11 is *IgL* (the immunoglobulin lambda gene).
3. t(2;8)(p12;q24): This occurs in about 5% of cases. The breakpoint is close to the distal end of c-*MYC*, and the partner gene on 2p12 is *IgK* (the immunoglobulin kappa gene).

Attention is drawn to the description of the location of the MYC breakpoint, as it has practical relevance for fluorescence *in situ* hybridization (FISH) studies. With some of the *MYC* probes currently available, the type of translocation dictates whether the signal stays on chromosome 8 or is relocated to the partner chromosome. Consequently, metaphases are needed for this type of study, as the abnormality is not evident in interphase cells. Other probes that span the *IgH* gene should be used for interphase FISH studies.

The prognosis associated with these abnormalities had been very poor in the past, but modern intensive treatments are increasingly more effective.

2.5. The t(9;22)(q34;q11)

The Philadelphia translocation (Ph), t(9;22)(q34.1;q11), which is found in >90% of cases of chronic myeloid leukemia (CML) and in about 3% of cases of acute myeloid leukemia (AML), also occurs in ALL. It is found in 2–5% of cases of childhood ALL, and in adults it has been shown by molecular studies to occur at a frequency directly proportional to age, such that by the age of 50 about 45% of ALLs may be Ph+ ***(12)***. For this reason, all adult cases of ALL should be screened for this abnormality by another technique (e.g., FISH or reverse transcription-polymerase chain reaction [RT-PCR]) if it is not found by a conventional cytogenetic study.

Part of an oncogene, c-*ABL* (mapped to 9q34.1), is translocated to a breakpoint cluster region (*BCR*) at 22q11. There are two common kinds of translocation, one in the "major" *BCR* which usually occurs in all cases of CML and some cases of ALL and codes for a 210-kDa hybrid phosphoprotein, the other in the "minor" *BCR* which usually occurs in ALL and produces a 185-kDa protein. The product in both cases has enhanced tyrosine kinase activity. Although in ALL the Ph usually disappears in remission (unlike the situation in CML), the prognosis is very poor with present chemotherapy protocols ***(13)***, including bone marrow transplant. It is often associated with other unfavorable factors, such as high white cell count and high incidence of extramedullary involvement (e.g., in the central nervous system), but a genetic study is the only reliable way of detecting this important abnormality.

In a few cases, the Ph has been found as part of a high hyperdiploid clone, and it seems possible that the detrimental effect on prognosis of the Ph may be less powerful than the beneficial effect of the hyperdiploidy ***(9)***; this tentative conclusion needs to be confirmed with more cases.

2.6. Abnormalities of 9p

Translocations and deletions involving 9p21 identify another cytogenetic subgroup, more common in older children, and clinically associated with bulky disease and hyperleukocytosis. Many, but not all, cases have T-cell phenotype. Some studies have reported a poor prognosis with a high incidence of extramedullary relapse ***(14)***. Two well established, unbalanced translocations involving 9p resulting in dicentric chromosomes are dic(9;12)(p11–13;p11–12) and dic(9;20)(p11;p11). The dic(9;12) is associated with B-cell precursor ALL, hypodiploidy, predominantly male sex, and a very good prognosis with low relapse rates ***(15)***. A common secondary abnormality is trisomy 8. *See* **Subheading 2.10.** for a description of the dic(9;20).

2.7. Abnormalities of 11q23

At least 50 different translocation partners for 11q23 are known ***(16)***, and the t(4;11)(q13;q23) has already been mentioned in the preceding. Some of these translocations are usually associated with other kinds of leukemia, but are occasionally found in ALL. For example, a t(9;11)(p21–22;q23) is usually associated with AML M5a, but several cases of ALL are known ***(17)***. The gene at 11q23 that is usually involved is variously known as *MLL, ALL1,* and *HRX*. It appears to be clinically important to determine whether or not this gene is involved; cases in which an abnormality of MLL is demonstrated generally have a worse prognosis than those without.

A t(11;19)(q23;p13) has been described as occurring in at least four conditions, namely acute monocytic leukemia, infantile B-lineage ALL, biphenotypic acute leukemia with high white cell count and poor prognosis, and T-cell ALL in older children, this last possibly having a relatively good prognosis. Attention is brought to this abnormality because it is particularly subtle and can easily be missed by conventional cytogenetics. There are two forms of this translocation, and each is best seen by a different kind of banding: t(11;19)(q23;p13.1) by G-banding and t(11;19)(q23;p13.3) by R-banding ***(18)***. If there is any suspicion that this abnormality may be present, a FISH study using a probe for the *MLL* gene is strongly advised.

2.8. Abnormalities of 12p, and the t(12;21)(p13;q22)

Visible rearrangements involving 12p12–13 by translocation or deletion occur in up to 10% of cases of childhood ALL. Several partner chromosomes are involved, two of the most common being t(7;12)(q11;p12) and the dic(9;12)(p11–12;p12) previously mentioned.

In up to 25% of cases of pre-B ALL, however, there is a cryptic translocation, t(12;21)(p13;q22), involving the *TEL* and *AML1* genes (more recently described as *ETV6* and *CBFA2*, respectively), which is very rarely suspected by conventional cytogenetics ***(19)***, as the material exchanged is morphologically almost identical. It has been associated with high remission rates and was thought to identify a good prognosis group. However, it is now known that there is a tendency to late relapse. Because patients with this abnormality therefore need particular long-term attention, and because it is so frequent, it is necessary to screen all cases of ALL by FISH or RT-PCR. FISH studies have revealed some other features associated with this translocation: in many cases part or all of the *TEL* gene on the apparently normal 12 has been deleted, a finding more common at diagnosis than in relapse. The t(12;21) can also be involved with other abnormalities. For example, a case with +21 may be shown to have a t(12;21) and a +der(21)t(12;21), and in the author's laboratory three out of four cases with what appeared to be a t(12;13)(p12;q14) have been shown by FISH to be t(12;21;13)(p13;q22;q14). Complex rearrangements involving a t(12;21) appear to be associated with a poorer prognosis. However, t(12;21) does not seem to co-occur with other common abnormalities in ALL, such as t(1;19), t(4;11), t(9;22) and high hyperdiploidy ***(20)***.

2.9. Chromosome 14

Abnormalities of chromosome 14 are common in ALL and two breakpoints are specifically associated with different phenotypes. At

14q11 are located the *TCR*-α and *TCR*-δ genes, and translocations at this locus are strongly associated with T-cell leukemia. However, at least two pre-B ALL cases are known ***(21)***. Clinically, patients with translocations at 14q11 often have a large tumor load, lymphadenopathy, involvement of extramedullary sites, and a poor prognosis. The other two known T-cell receptor sites, which also tend to be involved in T-cell ALL, are *TCR*-β at 7q34–36, and *TCR*-γ at 7p15.

In the laboratory now directed by the author of Chapter 6, Dr. Susana Raimondi, there has been a long history of particularly high success rates and clone detection rates in ALL ***(22)***, being well over 90% in most types ***(23)***. However, a lower clone frequency, little more than 60%, is found in T-cell cases ***(24)***.

The other commonly involved breakpoint is 14q32.3, the site of the *IgH* (immunoglobulin heavy chain) gene, and translocations with this breakpoint are usually associated with B-cell disorders, both ALL and lymphoma (*see* **Subheading 2.4.**).

Both breakpoints on the 14 are involved in inv(14)(q11q32), an inversion of the long arms of one chromosome, and a t(14;14)(q11;q32). The inversion is a fairly subtle abnormality that can be missed by an inexperienced cytogeneticist. Both of these abnormalities occur in T-cell tumors, and if this diagnosis is likely then a cytogenetic study should include analysis of divisions from cultures stimulated with a T-cell mitogen such as phytohemagglutinin (PHA). It seems likely that the 14q32 breakpoint in inv(14) and t(14;14) is outside the IgH locus which is involved in the translocations found in B-cell disorders.

2.10. Monosomy 20 and dic(9;20)(p11;q11)

An unusual but recurring abnormality, sometimes in the absence of any other abnormality, is apparent monosomy 20. If this is found, then further investigation is needed, as it has been found that most cases of –20 actually have a dicentric chromosome derived from the short arms of a chromosome 20 and the long arms of a chromosome 9, that is, dic(9;20)(p11;q11). Close inspection of the no. 9 chromosomes can show that one of them has unusual short arms, which are in fact the short arms of the missing chromosome 20.

2.11. Trisomy 21

Trisomy for chromosome 21 is a common constitutional abnormality (Down syndrome); this condition is associated with a perinatal leukemoid reaction that may resemble leukemia ***(25)***, and with an increased risk of developing acute leukemia, mostly megakaryocytic ***(26)***. *See* Chapter 12, **Subheading 5.1.2b.** Acquired trisomy 21 as the sole abnormality occurs in a wide variety of hematologic malignancies, and is the most frequent simple trisomy in childhood ALL. If this is detected in a patient with leukemia, two particular possibilities should be checked: (1) that it not constitutional; and (2) that it is not associated with a t(12;21)(p12;q22). Either trisomy or tetrasomy 21 is also almost always part of the pattern of gains in typical high hyperdiploidy.

3. Summary

Acute lymphoblastic leukemia is a fascinating disease for the cytogeneticist, as so many cases have a clone detectable by cytogenetics or FISH, and because identifying the abnormalities provides such useful information to the clinician. However, it is also a frustrating disease, as it has technical challenges such as a marked tendency for the sample to clot during harvesting, frequently poor chromosome morphology, and, especially in the high count cases, failure to provide any divisions at all for analysis. For these reasons, this book includes two chapters on the practical aspects of undertaking cytogenetic studies in ALL to illustrate contrasting approaches. The first is from a laboratory that is a world leader in its success rates, which has an enviably low sample/cytogeneticist ratio, and which is usually able to expect a good-sized sample commensurate with the importance given to a diagnostic cytogenetic study. The second is from a laboratory that also has a good success rate, despite having to cope with a higher workload and often much smaller samples. This is not to imply that each technique is limited to such circumstances; both are worthy of study and emulation.

References

1. Bennett J. M., Catovsky D., Daniel M. T., et al. (FAB Co-operative Group) (1981). The morphological classification of acute lymphoblastic leukemias: concordance amongst observers and clinical correlations. *Br. J. Haematol.* **47,** 553–561
2. Secker-Walker L. M., Lawler S. D., and Hardisty R. M. (1978). Prognostic implications of chromosomal findings in acute lymphoblastic leukemia at diagnosis. *Br. Med. J.* **23,** 237–244.
3. Gibbons B., MaCallum P., Watts E., et al. (1991) Near haploid acute lymphoblastic leukemia: seven new cases and a review of the literature. *Leukemia* **5,** 738–743.
4. Bloomfield C. D., Secker-Walker L. W., Goldman A. I., et al. (1989) Six-year follow-up of the significance of karyotype in acute lymphoblastic leukemia (The Sixth International Workshop on Chromosomes in Leukemia, London, 1987) *Cancer Genet. Cytogenet.* **40,** 171–185.
5. Heerema, N. A., Sather, H. N., Sensel, M. G., et al. (2000) Prognostic impact of trisomies of chromosomes 10, 17 and 5 among children with acute lymphoblastic leukemia and high hyperdiploidy (> 50 chromosomes). *J. Clin. Oncol.* **18,** 1876–1887.
6. Moorman A. V., Clark R., Farrell D. M., Hawkins J. M., Martineau M., and Secker-Walker L. M. (1995) Modelling chromosome gain in hyperdiploid ALL: a Leukemia Research Fund United Kingdom Cancer Cytogenetics Group Study. *Blood* **86,** 771a, Abstr. 3073.
7. Heim S., Alimena G., Billstrom R., et al. (1987) Tetraploid karyotype (92,XXYY) in two patients with acute lymphoblastic leukemia. *Cancer Genet. Cytogenet.* **29,** 129–133
8. Secker Walker L. M., Berger R., Fenaux P., et al. (1992) The prognostic significance of the balanced t(1;19) and unbalanced der(19)t(1;19) translocations in acute lymphoblastic leukemia. *Leukemia* **6,** 363–369.
9. Chessells J. M., Swansbury G. J., Reeves B., Bailey C. C., and Richards S. M. (1997) Cytogenetics and prognosis in childhood lymphoblastic leukemia: results of MRC UKALL X. *Br. J. Haematol.* **99,** 93–100.
10. Johansson, B., Moorman, A. V., Haas, O. A., et al. (1998) Hematologic malignancies with t(4;11)(q21;q23)—a cytogenetic, morphologic, immunophenotypic and clinical study of 183 cases. *Leukemia* **12,** 779–787.

11. Hayashi Y., Raimondi S. C., Look A. T., et al. (1990) Abnormalities of the long arm of chromosome 6 in childhood acute lymphoblastic leukemia. *Blood* **76,** 1626–1630.
12. Secker-Walker L. M., Craig J. M., Hawkins J. M., and Hoffbrand A. V. (1991) Philadelphia-positive acute lymphoblastic leukemia in adults: age distribution, BCR breakpoint and prognostic significance. *Leukemia* **5,** 196–199.
13. Fletcher J. A., Lynch E. A., Kimball V. M., Donnelly, M., Tantravahi R., and Sallan S. E. (1991) Translocation (9;22) is associated with extremely poor prognosis in intensively treated children with acute lymphoblastic leukemia. *Blood* **77,** 435–439.
14. Heerema, N. A., Sather, H. N., Sensel, M. G., et al. (1999) Association of chromosome arm 9p abnormalities with adverse risk in childhood acute lymphoblastic leukemia: a report from the Children's Cancer Group. *Blood* **94,** 537–44.
15. Mahmoud H., Carroll A. J., Behm F., et al. (1992) The non-random dic(9;12) translocation in acute lymphoblastic leukemia is associated with B-progenitor phenotype and an excellent prognosis. *Leukemia* **6,** 703–707.
16. Huret, J-L., Dessen, P., and Bernheim, A. (2001) An atlas on chromosomes in hematological malignancies. Example: 11q23 and MLL. *Leukemia* **15,** 987–989. (http://www.infobiogen.fr/services/chromcancer)
17. Swansbury G. J., Slater R., Bain B. J., Moorman A. V., and Secker-Walker L. M. (1998). Hematological malignancies with t(9;11)(p21–22;q23)—a laboratory and clinical study of 125 cases. *Leukemia* **12,** 792–800
18. Moorman A. V., Hagemeier A., Charrin C., Rieder H., and Secker-Walker L. M. (1998) The translocations t(11;19)(q23;p13.1) and t(11;19)(q23;p13.3): a cytogenetic and clinical profile of 53 patients. *Leukemia* **12,** 805–810.
19. Shurtleff S. A., Buijs A., Behm F. G., et al. (1995). TEL/AML1 fusion resulting from a cryptic t(12;21) is the most common genetic lesion in pediatric ALL and defines a subgroup of patients with an excellent prognosis. *Leukemia* **9,** 1985–1989
20. Raynaud, S. D., Dastugue, N., Zoccola, D., Shurtleff, S. A., Mathew, S., and Raimondi, S. C. (1999) Cytogenetic abnormalities associated with the t(12;21): a collaborative study of 169 children with t(12;21)-positive acute lymphoblastic leukemia. *Leukemia* **13,** 1325–1330.

21. Dube I. D. and Greenberg M. L. (1986) Phenotypic heterogeneity in three cases of lymphoid malignancy with chromosomal translocations in 14q11. *Cytogenet. Cell Genet.* **41,** 215–218.
22. Williams D. L., Harris A., Williams K. J., Brosius M. J., and Lemonds W. (1984) A direct bone marrow chromosome technique for acute lymphoblastic leukemia. *Cancer Genet. Cytogenet.* **13,** 239–257.
23. Williams D. L., Raimondi S., Rivera G., George S., Berard C. W., and Murphy S. B. (1985) Presence of clonal chromosome abnormalities in virtually all cases of acute lymphoblastic leukemia. *N. Engl. J. Med.* **313,** 640.
24. Heerema, N. A., Sather, H. N., Sensel, M. G., et al. (1998) Frequency and clinical significance of cytogenetic abnormalities in pediatric T-lineage acute lymphoblastic leukemia: a report from the Children's Cancer Group. *J. Clin. Oncol.* **16,** 1270–1278.
25. Brodeur G. M., Dahl G. V., Williams D. L., Tipton R. E., and Kalwinsky D. K. (1980) Transient leukemoid reaction and trisomy 21 mosaicism in a phenotypically normal newborn. *Blood* **55,** 69–73.
26. Zipursky A., Peeters M., and Poon A. (1987) Megakaryoblastic leukemia and Down's syndrome—a review. *Prog. Clin. Biol. Res.* **246,** 33–56.

6

Conventional Cytogenetic Techniques in the Diagnosis of Childhood Acute Lymphoblastic Leukemia

Susana C. Raimondi and Susan Mathew

1. Introduction

Cytogenetic analysis is an important aid in the classification of hematological disorders. Most types of leukemia display either numerical chromosomal abnormalities or structural rearrangements, mainly translocations. Nonrandom chromosomal abnormalities, which are being increasingly recognized, are especially useful in diagnosing the leukemic subtype and predicting treatment outcome. Molecular analysis of genes adjacent to the breakpoints of specific translocations and the study of the functions of their gene products have helped to clarify the complex interactions that promote leukemogenesis and perpetuate the leukemic phenotype *(1)*.

Acute lymphoblastic leukemia (ALL) is the most common childhood malignancy, comprising about 30% of all cases of pediatric neoplasia. ALLs can be classified into five subtypes based on the modal number of chromosomes: hyperdiploid with more than 50 chromosomes, hyperdiploid with 47–50 chromosomes, pseudodiploid (46 chromosomes with structural or numerical abnormalities), diploid (46 chromosomes), and hypodiploid (fewer than 46

From: *Methods in Molecular Biology, vol. 220: Cancer Cytogenetics: Methods and Protocols*
Edited by: John Swansbury © Humana Press Inc., Totowa, NJ

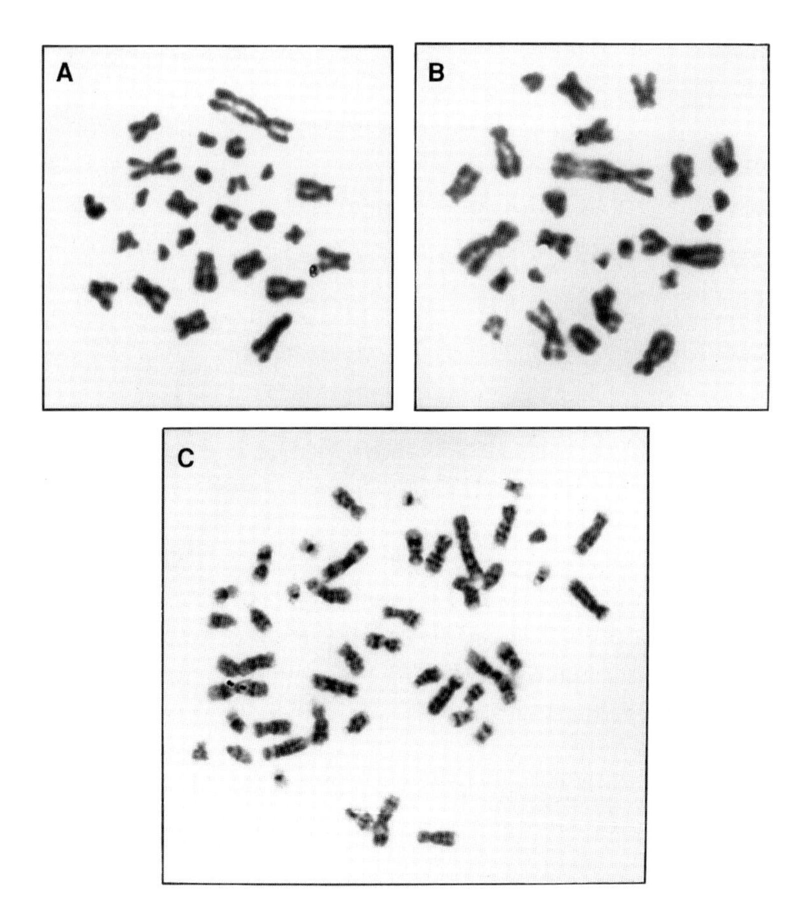

Fig. 1. Near haploid/normal metaphases in a child with ALL. Comparison of banding pattern of near haploid metaphases **(A,B)** with a normal metaphase **(C)**. The near haploid cell line contained 26 chromosomes and a large marker: add(1)(p36).

chromosomes). It is noteworthy to identify the patients with blasts containing a near-haploid modal number, which are infrequent but are associated with an extremely poor prognosis (**Fig. 1**). Recognition of ploidy as a distinctive cytogenetic feature in ALL has greatly enhanced our ability to predict treatment outcome *(2)*.

Defining ALL by the types of structural abnormalities found in the chromosomes of leukemic clones has led to impressive advances in understanding the biology of the disease and may suggest opportunities for risk-specific therapies *(3)*. **Table 1** describes the most common structural abnormalities found in ALL blasts and the proto-oncogenes associated with these abnormalities. The protein products of these proto-oncogenes may contribute to the disease process.

Most of the structural chromosome rearrangements found in ALL correlate with the leukemic cell immunophenotype. The t(8;14) and its variant translocations, t(2;8) and t(8;22), for example, were initially described in cases of Burkitt-type B-cell ALL with the French–American–British (FAB)-L3 morphology. In childhood ALL, illustrations of the close association between chromosomal translocation and phenotype of the leukemic cell include the t(1;19) in pre-B ALL, the t(4;11) and t(9;22) in B-cell precursor ALL, and the 7q35 and 14q11 rearrangements in T-cell ALL *(2)*.

Most nonrandom translocations associated with leukemias involve proto-oncogenes. The availability of molecular probes for sequences containing translocation breakpoints has greatly facilitated the development of molecular cytogenetics, including fluorescence *in situ* hybridization (FISH) and comparative genomic hybridization (CGH). These are described in the following chapters.

Conventional cytogenetic techniques performed on bone marrow aspirates or peripheral blood (when a high percentage of circulating blasts is present) yield karyotypes of the leukemic cells at the light microscopy level. Such cytogenetic findings are useful when characterizing the leukemic cells at diagnosis, identifying unique leukemic subtypes, correlating the chromosome findings with prognosis, monitoring the patient's response to chemotherapy, comparing the karyotype obtained at relapse with that at diagnosis, and directing therapy.

2. Materials

1. Centrifuge.
2. Incubator: 37°C.
3. Brightfield microscope.
4. Media: RPMI-1640 with L-glutamine (JRH Biosciences, Lenexa, KS).
5. Fetal bovine serum (FBS; Gemini-Bioproducts, Calabasas, CA).
6. Heparin: Preservative-free, 1000 U/mL (Fujisawa USA, Deerfield, IL).

Table 1
Recurrent Structural Chromosome Abnormalities in Childhood ALL

Abnormality	Percent in ALL overall	Percent in specific immunophenotype	Chromosome band/gene involved	
B-lineage ALL	3			
t(8;14)(q24.1;q32)	< 1	B-cell, 90	8q24.1/*MYC*	14q32/*IGH*
t(2;8)(p12;q24.1)	< 1	B-cell, 4–5	2p12/*IGK*	8q24.1/*MYC*
t(8;22)(q24.1;q11.2)	5–6	B-cell, 6–10	8q24.1/*MYC*	22q11.2/*IGL*
t(1;19)(q23;p13.3)	2–5	Pre-B, 90	1q23/*PBX1*	19p13.3/*E2A*
t(9;22)(q34;q11.2)	2	Early pre-B, 75	9q34/*ABL*	22q11.2/*BCR*
t(4;11)(q21;q23)	< 1	Early pre-B, 80	4q21/*AF4*	11q23/*MLL/ALL1*
t(5;14)(q31;q32)	< 1	Eosinophilia	5q31/*IL3*	14q32/*IGH*
t(17;19)(q22;p13.3)	< 1	Early pre-B	17q22/*HLF*	19p13.3/*E2A*
t(12;21)(p13;q22)[a]		Early pre-B, Pre-B	12p13/*TEL/ETV6*	21q22/*AML1/CBFA2*
T-lineage ALL	1			
t(11;14)(p13;q11.2)	< 1	T-cell, 7	14q11.2/*TCRA/D*	11p13/*RHOMB2/TTG2*
t(11;14)(p15;q11.2)	1	T-cell, 1	11p15/*RHOMB1/TTG1*	14q11.2/*TCRA/D*
t(10;14)(q24;q11)	< 1	T-cell, 5–10	10q24/*HOX11*	14q11.2/*TCRA/D*
t(8;14)(q24.1;q11.2)	< 1	T-cell, 2	8q24.1/*MYC*	14q11.2/*TCRA/D*
t(1;14)(p32;q11.2)	< 1	T-cell, 3[b]	1p32/*SCL/TCL5/TAL1*	14q11.2/*TCRA/D*
inv(14)(q11.2q32)	< 1		14q11.2/*TCRAD*	14q32/*IGH*
t(1;7)(p32;q35)	< 1		1p32/*SCL/TCL5/TAL1*	7q35/*TCRB*
t(1;7)(p34.1;q35)	< 1		1p34.1/*LCK*	7q35/*TCRB*

t(7;9)(q35;q34)	< 1	7q35/*TCRB*	9q34/*TAN1*
t(7;9)(q35;q32)	< 1	7q35/*TCRB*	9q32/*TAL2*
t(7;10)(q35;q24)	< 1	7q35/*TCRB*	10q24/*HOX11*
t(7;11)(q35;p13)	< 1	7q35/*TCRB*	11p13/*RHOMB2*
inv(7)(p15q35)	< 1	7p15/*TCRB*	7q35/*TCRB*
t(7;19)(q35;p13.3)		7q35/*TCRB*	19p13/*LYL1*
ALL of nonspecific lineage	4-3		
del(6q)	7–12		
t/del(9p)	3–5	9p22/p16/*MTS1*	
t/del(11q)	10–12	11q23/*MLL/ALL1*	
t/del(12p)		12p13/*TEL/ETV6*	

[a]*TEL/ETV6* gene rearrangements by FISH or RT-PCR positive for cryptic t(12;21) in 25% of cases with B-lineage immunophenotype.

[b]*TAL1* submicroscopic deletion in 15–26% of T-cells.

7. Colcemid: KaryoMAX, 10 μg/mL (Invitrogen, Carlsbad, CA).
8. Stainless steel wire: Diameter, 0.29-inch (Small Parts, Inc. Miami, FL), cat. no. SWX-029.
9. Wright's stain: Powder (Sigma, St. Louis, MO).
10. pHydrion buffers: In capsules, pH 7.00 ± 0.02 at 25°C (Micro Essential, Brooklyn, NY).
11. Trypsin: 0.25% (Invitrogen), stored as 5-mL aliquots in the freezer.
12. Hypotonic solution: Add 0.56 g of potassium chloride to 100 mL of deionized water. Make fresh solution before each experiment.
13. Carnoy's fixative (three parts methanol to one part of acetic acid): Combine 75 mL of methanol (high purity) and 25 mL of glacial acetic acid (high purity) in a 100-mL glass bottle. Make a fresh solution just prior to use.
14. Normal saline solution: Add 27 g of sodium chloride to 3 L of deionized water.
15. Stock buffer for Wright's stain: Add one pHydrion capsule to 100 mL of deionized water. Mix well and store at 4°C.
16. Working buffer for Wright's stain: Add 5 mL of stock buffer to 95 mL of deionized water.
17. Wright's stain stock solution: Place 100 mL of methanol in a beaker on a stirplate. While stirring at medium-high speed, gradually add 0.3 g of powdered Wright's stain. Cover beaker to prevent splashing and stir for at least 30 min. Filter the solution through double Whatman no. 40 paper and store at 4°C in a brown bottle.
18. Working stain: Add 10 mL of Wright's stain stock solution to 40 mL of working buffer.

3. Methods

3.1. Collecting the Specimen

Use aseptic technique. It is extremely important to collect the bone marrow sample for cytogenetic evaluation immediately following the sample that is collected for morphologic studies (aspirate smears). This early bone marrow aspirate (second syringe) will ensure that a concentrated marrow sample is obtained. The best results are obtained when the sample is received in the laboratory within 30 min to start processing.

3.1.2. Bone Marrow

1. Put 15–20 mL of RPMI-1640 media with 15% FBS and 0.05 mL of heparin solution into a 50-mL centrifuge tube.
2. Collect the sample into the prepared tube; each chromosome analysis requires 1–2 mL of bone marrow aspirate. To prevent clotting, immediately cap the tube and mix by inverting several times.
3. Preparing the sample: In the laboratory, split the collected marrow sample into two or more 15-mL centrifuge tubes. Each tube should contain no more than 0.5 mL of bone marrow aspirate. When patients have a very high white blood cell count (>70,000/mm^3), less marrow aspirate should be added per tube: 0.1–0.3 mL. Therefore, the number of tubes prepared should be adjusted according to the amount of bone marrow collected in the 50-mL tube. One of the tubes is processed immediately; the other is incubated at 37°C overnight. Both tubes are processed as described in **Subheading 3.2.**

3.1.2. Peripheral Blood

Peripheral blood cultures should be performed only when circulating blasts are present (preferably >25%).

1. Draw 5–10 mL of blood into a syringe coated with preservative-free heparin and transfer to a 15-mL centrifuge tube. Centrifuge at 200*g* for 6 min.
2. Remove buffy coat/plasma and transfer to two to four 15-mL centrifuge tubes, each of which contains 9–10 mL of RPMI-1640 with 15% FBS and 0.05 mL of heparin solution. The amount of buffy coat/plasma put in each tube varies according to the white blood cell count of the patient's sample—if <30,000/mm^3, add 1 mL of buffy coat/plasma; if 30,000–70,000, add 0.5 mL; if 70,000–150,000, add 0.2 mL; and if >150,000 add 0.1 mL.
3. Incubate overnight at 37°C and process as described in **Subheading 3.2.**

3.2. Processing the Specimen

1. Add 0.05–0.1 mL of colcemid to each centrifuge tube. Recap the tube and invert several times to mix. Let it stand at room temperature for 25 min.
2. Centrifuge at 200*g* for 10 min.

3. Remove the supernatant, leaving approx 0.25 mL above the cell pellet. Resuspend the cell pellet with a stirring wire or a Pasteur pipet *(4)*.
4. While stirring, add five drops of hypotonic solution. Then slowly add an additional 10 mL of hypotonic solution while stirring and mix well. Recap the tube and let it stand at room temperature for 25–30 min.
5. Centrifuge tube at 200*g* for 6 min. After centrifugation, there will be a layer of white cells above the red cells.
6. Remove most of the supernatant, leaving approx 0.25 mL above the cell pellet.
7. Resuspend the cell pellet. While stirring, add five drops of 3:1 Carnoy's fixative to each tube. This step is important to prevent clumping of cells. Then add an additional 10 mL of Carnoy's fixative and resuspend. Recap the tube and keep for 15 min at room temperature.
8. Centrifuge tube at 200*g* for 6 min.
9. Remove the supernatant and repeat **step 7**. Let the tube stand at room temperature for 10 minutes.
10. Centrifuge tube at 200*g* for 6 min.
11. Remove supernatant. Repeat changing the Carnoy's fixative, until the cell pellet is white. Prior to slide making, resuspend the cell pellet. Add fresh 3:1 Carnoy's fixative a few drops at a time until the final suspension is slightly cloudy.

3.3. Making Slides

For best results, adjust the humidity of the area to between 50% and 75%.

3.3.1. Hotplate Method

1. Slides are precleaned in 75% alcohol and refrigerated in deionized water.
2. Shake off excess water.
3. Aspirate the final cell suspension into a siliconized disposable glass or plastic Pasteur pipet with rubber bulb.
4. Hold the filled pipet 6–12 inches above the slide.
5. Tilt the slide at a 45° angle to the floor.
6. Release one to two drops of the cell suspension onto the slide. The drop should land near the frosted end of the slide. Very gently, blow once or twice on the slide.

7. Wipe the bottom of the slide with gauze and place on a warm (>30°C) or hot (60°–75°C) plate until the slide is dry.
8. Etch the accession number onto each slide.
9. Examine a test slide under a phase-contrast microscope to ensure that the sample has metaphase chromosomes (*see* **Note 1**).

3.3.2. Flame Method

The flame method, using an alcohol burner, is applied only when the metaphases are poorly spread *(4)*. Although at present rarely used, this method may help to spread metaphases that do not respond to other methodologies.

1. Follow **steps 1–5** in **Subheading 3.3.**
2. With the slide tilted, release one drop of the cell suspension onto the slide. The drop should land on the bottom half of the slide.
3. Flame the slide holding the slide at a 45° angle. Move the slide in the flame for approx 2 s. Place slides in drying racks and let them dry completely.
4. Etch each slide with the accession number.
5. Examine a test slide under a phase-contrast microscope to ensure that chromosomes are adequately spread.

3.4. Aging the Slide

Optimal aging, an important step in successful chromosome banding for hematologic disorders, can be achieved by either natural or rapid methods.

3.4.1. Natural Aging

The slides are aged by leaving them at room temperature. The time varies (3–10 d) and, on rare occasions, good G-banding may be obtained immediately after harvest.

3.4.2. Rapid Aging

Rapid aging can be achieved either by placing the slides in a conventional oven at 75–90°C for 10–30 min or by microwaving the slides at the high setting for 2–5 min. Rapid aging leads to adequately banded metaphases for analysis.

3.5. G-Banding Technique (Using Wright's Stain)

1. Line up five Coplin jars and fill the first with 5 mL of trypsin plus 45 mL of normal saline solution, the second with 50 mL of normal saline, the third with 10 mL of Wright's stain stock solution and 40 mL of buffer for Wright's stain, and the last two with 50 mL of deionized water.
2. Process the slides one at a time through the banding setup until results are optimized. For fresh slides, start with 5–15 s in the trypsin solution, rinse in saline by dipping the slide two or three times, stain for 1–2 min, and rinse twice in deionized water. View and adjust times as needed. Let slides air-dry.

4. Note

1. When no metaphases are available in the diagnostic sample, the attending physician is contacted to discuss the possibility of obtaining another bone marrow aspirate, or a peripheral blood sample if > 25% blasts are present.

References

1. Rabbitts, T. H. (1991) Translocations, master genes, and differences between the origins of acute and chronic leukemias. *Cell* **67,** 641–644.
2. Raimondi, S. C. (1993) Current status of cytogenetic research in childhood acute lymphoblastic leukemia. *Blood* **81,** 2237–2251.
3. Pui, C-H. (1995) Childhood leukemias. *N. Engl. J. Med.* **332,** 1618–1630.
4. Williams, D. L., Harris, A., Williams, K. J., Brosius, M. J., and Lemonds, W. (1984) A direct bone marrow chromosome technique for acute lymphoblastic leukemia. *Cancer Genet. Cytogenet.* **13,** 239–257.

7

Chromosome Preparations from Bone Marrow in Acute Lymphoblastic Leukemia

Cytogenetic Techniques

Ann Watmore

1. Introduction

It is well recognized that malignant cells from acute lymphoblastic leukemias (ALLs) generally yield less satisfactory metaphases for analysis than cells from other diseases when standard laboratory processes are applied, in spite of the fact that karyotypically normal cells from the same sample can produce high-quality divisions under the same conditions. Thus it is of primary importance in the analysis of ALL preparations not simply to select the best available metaphases for analysis, or the malignant clone may not be sampled. In addition to this, some variation to routine procedures may enhance the quality and relative quantity of abnormal metaphases and facilitate accurate analysis.

As with all laboratory procedures, there are elements that are not themselves critical to the process but that may contribute to optimizing results in a particular laboratory. Indeed variation of several such elements may interact to produce success or failure. It is necessary to experiment with these noncritical elements, for example,

From: *Methods in Molecular Biology, vol. 220: Cancer Cytogenetics: Methods and Protocols*
Edited by: John Swansbury © Humana Press Inc., Totowa, NJ

choice of medium, culture vessel, and so forth, when developing techniques suitable for ALL, as well as taking into account points that may be of more specific importance, for example, cell density. Notes reflecting local experience are collected at the end of the protocol.

2. Materials

1. Transport medium: 100 mL of McCoy's 5A medium with 2.5 mL of *N*-2-hydroxyethylpiperazine-*N*'-2-ethanesulfonic acid (HEPES) buffer, 1 mL of penicillin (5000 IU/mL); 1 mL of streptomycin (5000 µg/mL), 200 U of preservative-free heparin.
2. Culture medium: 100 mL of McCoy's 5A medium with 2.5 mL of HEPES buffer (*see* **Note 1**); 1 mL of penicillin (5000 IU/mL); 1 mL of streptomycin (5000 µg/mL); 10 mL of pooled human serum (*see* **Note 2**).
3. Colcemid: Working solution: 5 µg/mL, that is, stock solution of 10 µg/mL diluted twice; final concentration, 0.05 µg/mL.
4. Hypotonic solution: 0.075 *M* potassium chloride (KCl, 5.5 g/L in distilled deionized water).
5. Fluorodeoxyuridine (FdUr): Working solution, 10^{-5} *M*; final concentration, 10^{-7} *M*. (**Note:** *FdUr is cytotoxic and appropriate safety procedures should be followed.*)
6. Uridine: working solution, 4×10^{-4} *M*; final concentration, 4×10^{-6} *M*.
7. Thymidine: working solution, 10^{-3} *M*; final concentration, 10^{-5} *M*.
8. Fixative: Analytical grade methanol–glacial acetic acid (3:1), prepared immediately before use on each occasion as required.

   All working solutions should be prepared and used under sterile conditions and stored at 4°C; two exceptions are KCl, which is stored at 37°C for use and does not need to be kept sterile, and fixative, which is made up immediately before use.
9. Microscope slides: High-quality, precleaned glass slides.

3. Methods

3.1. Transport of Samples

Bone marrow may be transported without medium, in which case a heparinized container must be used.

3.1.1. Samples for Culture

Place about 1 mL of bone marrow aspirate directly into 5–10 mL of transport medium in a sterile universal container. Rapid transportation is recommended although samples delayed for up to 24 h, for example, in the mail, can give adequate results.

3.1.2. Samples for True Direct Processing

Place one drop of aspirate from the syringe directly into 10 mL of transport medium containing 0.05 μg/mL of colcemid. These samples should reach the laboratory as soon as possible and preferably within 2 h.

3.1.3. Supply of Transport Medium

Transport medium may be made up by the lab and dispatched to users as required, marked with a 1-month expiration time and instruction to maintain sterility.

3.2. Culturing

3.2.1. Setting Up Cultures

Sterility must be maintained throughout and appropriate safety regulations adhered to. The choice of culture vessel will reflect what is routinely available, although conical bottomed vessels are not recommended. This laboratory currently achieves its best results in flat-bottomed glass universal containers.

Under sterile conditions, add an appropriate amount of sample to achieve a final cell density of about 10^5 nucleated cells/mL in a sterile glass universal container (*see* **Note 3**). Make up to 10 mL with culture medium (cell counting protocols are given elsewhere but with experience a suitable cell density can be judged by eye.)

3.2.2. Recommended Cultures

There is a greater chance of obtaining abnormal metaphases if bone marrow is not cultured before metaphase arrest. Placing the sample directly into medium containing colcemid followed by quick transportation and processing is recommended. If abnormal metaphases are not present after this, it is very unlikely that they will be seen after culture. Although this may be the most reliable way of sampling a malignant clone, it will not necessarily yield the best quality metaphases. This laboratory's experience is that no one particular regimen consistently yields good quality divisions and a variety of cultures should always be attempted. Each of the following cultures has different theoretical advantages which should be considered when selecting which one(s) to use:

True direct
Same day
Short-term unsynchronized
Synchronized
Overnight colcemid

A reasonable approach is to use a true direct if available along with short-term unsynchronized and synchronized cultures. Unsynchronized cultures of 48 h may be harvested if earlier cultures are not adequate.

Same day cultures appear to have no advantage in terms of quality and do not always show up the malignant clone if the sample is more than 1–2 h old; however, if the clone is evident a quick result can sometimes be achieved this way. Overnight colcemid-treated cultures often produce relatively small chromosomes but can be useful if first cultures indicate an unreasonably low mitotic index, or if preparations for metaphase *in situ* hybridization techniques are required.

Most time intervals given have been designed to fit in with laboratory routine and may be altered. Increasing the period of metaphase arrest with colcemid should give a greater number of divisions although chromosomes will contract if exposed to colcemid for very long periods, for example, overnight. Here the advantage of a higher mitotic index may be outweighed by a substantial proportion of the cells being unanalyzable. The schedule for FdUr synchronization is less flexible. It is usual to "block" cell division overnight but longer periods are not harmful. When the "block" has been released it is assumed that cells will be in division some 5–7 h later, although neither the efficacy of blocking nor the cell cycle time are as predictable as they are for phytohemagglutinin (PHA)-stimulated lymphocytes.

3.2.2.1. TRUE DIRECT. Harvest immediately on arrival in the laboratory (*see* **Subheading 3.3.**).

3.2.2.2. SAME DAY CULTURE.

1. Set up culture as described in **Subheading 3.2.1.**
2. Add colcemid at 0.05 µg/mL.
3. Incubate at 37°C for 1–6 h.
4. Harvest (*see* **Subheading 3.3.**)

3.2.2.3. SHORT-TERM UNSYNCHRONIZED CULTURE.

1. Set up culture as described in **Subheading 3.2.1.**
2. Incubate at 37°C for 18–24 h (and/or 48 h).
3. Add colcemid at 0.05 µg/mL and return to the incubator for approx 5 h.
4. Harvest (*see* **Subheading 3.3.**).

3.2.2.4. FDUR SYNCHRONIZED CULTURES. Malignant cells are not effectively synchronized by the "block and release" techniques used for stimulated lymphocyte cultures. Nonetheless, these procedures may result in better quality metaphases than in other cultures and some form of synchronization is worth applying as an additional regimen.

1. Set up culture as described in **Subheading 3.2.1.**
2. Add FdUr at a final concentration of 10^{-7} *M* and uridine at 4×10^{-6} *M*.
3. Incubate at 37°C for about 18 h (or up to 42 h if more convenient).
4. Add thymidine at a final concentration of 10^{-5} *M* to release the FdUr "block."
5. Return to incubator for 4.5–5.5 h.
6. Add colcemid at 0.05 µg/mL and return to incubator for 1 h.
7. Harvest (*see* **Subheading 3.3.**)

3.2.2.5. OVERNIGHT COLCEMID-TREATED CULTURE.

1. Set up culture as described in **Subheading 3.2.1.**
2. Add colcemid at 0.05 µg/mL.

3. Incubate overnight.
4. Harvest (*see* **Subheading 3.3.**)

3.3. Harvesting Procedure

1. After colcemid treatment, transfer the culture to a clear, conical-bottomed centrifuge tube and centrifuge at 240*g* for 10 min.
2. Remove supernatant medium and resuspend the cell pellet thoroughly.
3. Add 0.5 mL of prewarmed KCl and mix (*see* **Note 4**).
4. Incubate at 37°C for 10 min.
5. Centrifuge at 240*g* for 5 min.
6. Carefully remove most of the supernatant and resuspend the pellet thoroughly.
7. Add fixative dropwise with continuous vigorous agitation to maintain cells in free suspension during fixation. This should be a quick and smooth process until 1–2 mL of fixative have been added. Add fix more liberally but still with agitation up to 5 mL as necessary (*see* **Note 5**).
8. Centrifuge at 240*g* for 10 min.
9. Remove the supernatant fixative, resuspend the pellet, and add fresh fixative, continuing to mix.
10. Repeat **steps 8** and **9** at least once, or until any brown coloration has disappeared.
11. After the final centrifugation, add only enough fixative for slide preparation (**Subheading 3.4.**). Alternatively, add more fixative, cap the tube, and store at –20°C until required.

3.4. Slide Preparation and Chromosome Banding

For all cytogenetic preparations, slide making is a critical step, and time invested in perfecting this skill is essential. In experimental work, accurate assessment of variation in results achieved with different culture regimens is often hampered by the internal variation between paired slides being greater than any variation between regimens. Several attempts at slide making may be necessary before optimum spreading is achieved (*see* **Note 6**).

1. Soak microscope slides in Decon® detergent solution overnight (*see* **Note 7**).

2. Wash slides individually under a hot tap, rinse in cold water, and store in distilled deionized water until use.
3. Resuspend the cell pellet in enough fresh fixative to produce a "just cloudy" suspension. If cultures have been stored at –20°C they should be centrifuged and fresh fixative added before slides are prepared.
4. Take up some cell suspension in a Pasteur pipet in one hand and hold a slide in the other hand with forceps. Drain most of the water from the slide (e.g., onto blotting paper).
5. Holding the slide horizontally, drop a single drop of suspension centrally from just above the slide.
6. Quickly and carefully place the slide on a level surface and blot off the water that has moved to the ends of the slide.
7. Allow to air-dry at room temperature.
8. Once aged (*see* **Note 8**), band with 0.1% trypsin for a few (<10) s (*see* **Note 9**) and stain with Leishman's stain for 1.5-2 min.

4. Notes

1. A variety of basal media may be suitable other than McCoy's 5A, for example, Ham's FI0 or RPMI 1640. TC199 as used for stimulated lymphocyte culture is not recommended.
2. Fetal calf serum is most commonly used in culture media but better results have been obtained in this laboratory with pooled human serum. This must be sterile filtered and has a shelf life of approx 3–4 wk at –20°C. Commercially prepared human AB serum is an alternative.
3. Conventional techniques suggest a cell density of about 10^6 nucleated cells/mL. However, bone marrow from patients with ALL is often very cellular, and handling difficulties experienced during harvesting may be related to this. Good results have been obtained with cell densities of about 10^5/mL and difficulty in fixation is not experienced with such less dense cultures.
4. The volume of hypotonic solution used in many standard procedures is as high as 10 mL. Experimentation in this laboratory has shown that even for routine purposes 1.5 mL of hypotonic is sufficient to treat a 10-mL culture without loss of chromosome spreading and often the mitotic yield is higher than when greater volumes are used. Even smaller volumes (0.5 mL) are preferred for ALL cultures as they are more dilute than others.

5. Fixation is most critical and is the stage where problems may occur. The following notes may help in avoiding difficulties or rescuing problem cases.
   a. Fixative: This must be made freshly within minutes of use.
   b. Loose cell pellets: If cells do not compact well on centrifugation but trail to the surface, the culture can be fixed without removing the hypotonic solution. **Steps 8** and **9** should be repeated more times to ensure complete fixation. Cultures behaving this way may form clumps when fixative is added. This will be worsened if the addition of fixative is too rapid or the culture is not continuously agitated with vigor.
   c. Interference from red blood cells: an over abundance of red blood cells may lead to poor fixation with a brownish gelatinous precipitate in spite of adequate agitation. This material will prevent effective spreading during slide making and may be removed by adding 5–10 mL of distilled water to the cell pellet and mixing thoroughly, centrifuging, and repeating if necessary. Several fixation steps will then be required.
6. There are many alternative methods of producing suitably spread metaphases, and different laboratories will produce equivalent results in quite different ways. To optimize spreading in ALL, it is necessary to be aware of the variables which affect slide quality within the laboratory such as slide temperature, density of cell suspension, number of drops used, degree to which the slide is drained, and the temperature and humidity of the laboratory. There is no substitute for experimentation into the interaction of these elements to provide a basis for good slide preparation. Even so, individual cases of ALL may require repeated sessions of slide preparation before the best results are achieved.
7. Slides: High-quality, precleaned slides are essential to give best results. They should still be cleaned to remove any traces of grease before use. Good quality metaphases will spread to a sufficient extent for analysis on poor slides but inferior ones such as those in ALL may not.
8. Aging slides for banding: More reproducible banding can be achieved if slides have aged. They may be left to age naturally for several days or aged more rapidly by one of the following two methods:
   a. After leaving overnight at room temperature, incubate slide at 80°C for about 1 h.

   b. Stain with Leishman's stain. Remove stain with fixative and rinse with ethanol or methanol. Air-dry. Place on a hotplate at 100°C for 10–15 min.

9. As for all cytogenetic preparations, the quality of banding is related to the morphology of the chromosomes. In general, however, the poorer morphology chromosomes often encountered on some types of ALL require shorter trypsin treatment. As little as 4 s is effective in this laboratory.

8

Lymphoid Disorders Other than Common Acute Lymphoblastic Leukemia

Background

John Swansbury

1. Introduction

This chapter describes the background to cytogenetic studies in lymphoid disorders other than common acute lymphoblastic leukemia (ALL); these include the less primitive kinds of ALL, such as T-cell ALL and B-cell ALL, all kinds of lymphoma, the chronic lymphoproliferative disorders, and malignant myeloma. Most of these disorders are not as amenable to cytogenetic study as the acute leukemias and chronic myeloid leukemia; the number of published cases is correspondingly fewer, although still greater, so far, than those on solid tumors. It is often not so easy to obtain a sample of the primary malignant tissue, such as lymph node, and the cytogeneticist may have to make do with other, infiltrated tissue, such as blood or bone marrow. Although these can be successful for some types of malignancy, for other types the success rate is very low.

Under normal circumstances, circulating lymphoid cells are mature cells that would not divide spontaneously but only in response to the presence of an antigen. Consequently, the cytogeneticist will need to

From: *Methods in Molecular Biology, vol. 220: Cancer Cytogenetics: Methods and Protocols*
Edited by: John Swansbury © Humana Press Inc., Totowa, NJ

use reagents (mitogens) to try to stimulate them into division. It is part of its natural immune response that a lymphoid cell is transformed into a cell that is capable of dividing. However, this immune response may be stunted by the disease, as the malignant cells may accumulate at an immature stage that is incapable of the normal response to an antigen: Immature T cells lack the TCR/CD3 complex in the cell membrane and immature B cells lack the Ig/BCR complex, so the response to all antigens may be weak or absent. Furthermore, if treatment has been started then the response may be severely inhibited by the immunosuppressive effect of the treatment itself.

Fortunately, some immature malignant lymphoid cells in the circulation may still be at a stage capable of spontaneous division, although at a low rate, and these can be harvested from short-term cultures without mitogens. Consequently, it is important that a variety of both stimulated and unstimulated cultures are set up. Although this greatly increases the amount of laboratory work, plus the amount of screening and analysis required, attempts to make do with a more limited range of cultures will tend to result in a higher incidence of failure to find a clone. The relatively large amount of work involved in a proper cytogenetic study of lymphoid disorders can place a strain on the manpower and finances of a Cytogenetics Unit, an imposition that is rarely appreciated by the referring clinician.

Despite all the extra efforts, conventional cytogenetic studies may not find any abnormality, especially in malignant myeloma (MM) and chronic lymphocytic leukemia (CLL), which are notoriously difficult. This is attributable mainly to two common problems: Often the mitotic index is very low, so there are few divisions of any kind, and/or most of the divisions present are cytogenetically normal and so probably do not derive from the malignant cell population. For this reason, there has been a tendency in some centers to dispense entirely with a conventional cytogenetic study and proceed directly to using fluorescence *in situ* hybridization (FISH), screening for the most significant abnormalities. This policy is understandable but disappointing, as it limits the diagnostic service to using what is already known and is unlikely to extend our knowledge of the genetics of these disorders much further.

Most lymphoid cells have characteristics of one of two lineages, described as being of T (thymus) or B (bursa) type. These types generally respond to different mitogens, and are associated with different diseases and different cytogenetic abnormalities. Malignant cells that arise at a primitive stage of cell development may not fall into either class; it is usually these that tend to give rise to ALL—the "common" ALL immunophenotype is non-T-non-B, or pre-B and early pre-B. It is the lymphoid cells at further stages of differentiation that give rise to B-ALL, Burkitt lymphoma, T-ALL and T-lymphoblastic lymphoma, and so forth. There is a wide variety of malignancies arising from lymphoid cells, and some of these are listed in **Table 1**.

It is therefore important for the cytogeneticist to know the immunophenotype of the patient's disease, in advance, so that the appropriate T-cell or B-cell mitogens can be used in setting up stimulated cultures. However, if this information is unavailable at the time of receiving the sample, cultures for both types of cells should be set up.

Most B-cell mitogens can also stimulate T cells, so the divisions obtained from a mitogen-stimulated culture should not be assumed to be from the disease lineage unless they have a clonal abnormality. Because the normal cells may have a greater immune response, and the malignant cells usually have a low mitotic index, it may be necessary to analyze a relatively large number of divisions to improve the chances of detecting a clone. In some patients, all the divisions will be found to be abnormal; the incidence of clones varies greatly according to the type of malignancy being studied, as described in more detail later.

The clinical and biological backgrounds of four types of lymphoid malignancies are described in more detail in the following subheadings:

2. The Lymphomas

There have been many proposed systems for the classification of lymphomas; these originated not simply for academic interest but

Table 1
Typical Malignancies of Lymphoid Cells

Primitive non-T-non-B malignancies	
Common, null, and pre-B acute lymphoblastic leukemia (ALL)	
Immature B-cell malignancies	**Immature T-cell malignancies**
B-ALL	T-ALL
Burkitt lymphoma	
Mature B-cell malignancies	**Mature T-cell malignancies**
B-chronic lymphocytic leukemia (CLL)	T-CLL
B-prolymphocytic leukemia (PLL)	T-PLL
Non-Hodgkin's lymphoma (NHL)	Adult T-cell leukemia/lymphoma (ATLL)
Hairy cell leukemia (HCL)	Cutaneous T-cell lymphomas (CTCL), namely Mycosis fungoides and Sezary syndrome
Waldenstrom's macroglobulinaemia (WM)	Peripheral T-cell lymphomas
Malignant myeloma (MM)	Large granular lymphocytic leukemia
Plasma cell leukemia	
Splenic lymphoma with villous lymphocytes (SLVL)	
Hodgkin's disease of B-cell type	

rather because of the association between histological appearance and survival, which can range from a few months up to many years. The first widely accepted system, that of Rapapport, appeared in 1966 and established, for example, that patients with lymphomas comprised of small cells collected into nodes (or follicles) generally had a long survival, and those with large cells in a diffuse arrangement had a short survival. Other classifications were devised that extended these associations, and in 1982 a Working Formulation was drawn up to try to combine all the classifications into a system that would have international recognition. However, this compromise, based almost entirely on a morphologic description of the cells, greatly restricted any further development. The Kiel classification, updated in 1988, had more scope and used immunologic data for primary divisions and cell morphology for subdivisions, and linked these to the concept of high-grade and low-grade lymphomas. Further advances in understanding the biology and clinical characteristics of lymphoma were used to draw up the new Revised European–American Lymphoma (REAL) classification in 1994, in which the definition of each type of lymphoma incorporated data from morphology, immunology, genetics, location of origin, and clinical characteristics. At the same time the international prognostic index (IPI) was defined, in which clinical and biological features are used to calculate a prognostic score, such that patients with an IPI of 4 or more will have a median survival of about 18 mo, but those with an IPI of 3 or less will have a median survival of over 7 y. The further development of lymphoma classifications is being continued by the World Health Organization, in conjunction with its classification of other hematologic malignancies.

The variety of classifications in use during the last two decades has hampered attempts to correlate chromosome abnormalities with lymphoma types. The situation is complicated further by the disease in some patients evolving from one kind to another. In such cases, it is unclear whether the clonal abnormalities found relate to first disease subtype or the second. In a few cases, cytogenetic abnormalities that are associated with different kinds of lymphoma can be found to occur together in the same clone, which is thought to be further evidence for disease evolution through different types of lymphoma. There are several strong cytogenetic/diagnostic associations, but the occasional appearance of certain abnormalities in unexpected lymphoma types raises the question of whether (1) the diagnosis was incorrect, (2) the classification system was inadequate, (3) the disease had transformed from a different type, or (4) the patient's disease was truly exceptional.

Lastly, lymphomas are related to ALL and some cytogenetic abnormalities occur in both types of disease, especially a del(6)(q). However, even some of the classic abnormalities that are particularly associated with ALL, such as t(1;19)(q23;p13) and t(4;11)(q21;q23), have been occasionally reported in non-Hodgkin lymphoma.

2.1. Recurrent Cytogenetic Abnormalities Found in Lymphomas

A few cytogenetic abnormalities occur in almost all kinds of lymphoma, and these include translocations involving chromosome 1, deletion of part of the long arms of chromosome 6 (especially involving bands q21 and q27), trisomy 12, and, particularly in B-cell lymphomas, translocations to band 14q32. Compared to the clones found in many leukemias, those seen in lymphomas are more often complex. Fortunately for the cytogeneticist, however, the chromosome morphology tends to be generally slightly better. A complex clone has often been associated with a poor prognosis, irrespective of the particular abnormalities present.

A long established type of lymphoma is Hodgkin's disease. All other lymphomas are grouped under the term non-Hodgkin lymphoma (NHL). These are mostly B-cell type, and are subdivided according to their cell size (small, intermediate, large), whether or not they are cleaved, how the cells are distributed (follicular, diffuse), which organs are involved, and/or whether or not the bone marrow is infiltrated (high grade or low grade). Some combinations of these criteria are much more common than others, and some have specific names. Examples of those that are of particular interest to cytogeneticists are listed below, in alphabetic order.

Adult T-cell leukemia-lymphoma (ATLL): Trisomy for chromosome 3 is common in this type of lymphoma, as are the abnormalities involving the *TCR* genes (*see* "T cell lymphomas").

Anaplastic large-cell lymphoma (ALCL): This has a favorable prognosis, a particular immunophenotype (Ki1+), and has been associated with t(2;5)(p23;q35), which fuses the nucleophosmin gene on 5q35 with the anaplastic lymphoma kinase gene at 2p23. The association is not absolute: No t(2;5) has been detected in some cases of ALCL, and the translocation has been described in other kinds of lymphoma ***(1)***. However, it is likely that the translocation defines the disease more precisely than morphologic or immunologic methods ***(2)***. Up to 90% of cases do not have bone marrow involvement.

Burkitt lymphoma (BL) is a subgroup of diffuse, small noncleaved cell, high-grade lymphoma. Cytogenetically it is indistinguishable from the L3 French–American–British (FAB) type of acute lymphoblastic leukemia (ALL L3); almost all cases have one of three translocations, t(8;14)(q24;q32), t(2;8)(p12;q24), and t(8;22)(q24;q11), all involving a breakpoint at 8q24, (*see* Chapter 5), the locus of the *c-Mmc* gene.

Diffuse large cell lymphoma (DLBCL), including immunoblastic lymphoma: These constitute 85% of intermediate-grade lymphomas. In about 40% of cases, these lymphomas appear outside lymph nodes, for example, in the digestive tract, skin, bone, thyroid, and testes. The most common cytogenetic abnormalities are t(3;22)(q27;q11), t(2;3)(p12;q27), and t(3;14)(q27;q32) (all of which involve the *BCLG* gene), a t(14;15)(q32;q11–13) involving the *BCL8* gene on 15q ***(3)***, and abnormalities of 1q21. This last class of abnormality is associated with a poor prognosis ***(4)***. The *BCL6* gene has also been described in translocations with a dozen other partners.

Diffuse mixed, small cleaved and large cell lymphoma: Abnormalities of 19q11-q13 have been associated with this type.

Diffuse, small lymphocytic, low-grade, lymphoma: A subset of this type is identified by t(11;18)(q21;q21) as the sole abnormality, and this disease usually has an indolent course.

Follicular, small cleaved cell, low-grade lymphoma (FL) is the most common type of lymphoma in adults; it represents 30% of all

NHLs, and 65% of all low-grade NHLs *(5)*. However, it is rare in children. The typical cytogenetic abnormality is a t(14;18)(q32;q24), which involves the *IgH* gene at 14q32 and the *BCL2* gene at 18q21 *(6)*. When this translocation occurs in other NHLs, it is thought that these may have evolved from FL. Less common variants are t(2;18)(p12;q21) fusing the *IgK* and *BCL2* genes, and t(18;22)(q21;q11), fusing the *BCL2* and *IgL* genes. Most clones are complex, and it has not yet been possible to discover the clinical or biological significance of the abnormalities that are secondary to the t(14;18) *(7)*. The most common secondary abnormality is gain of an extra 18 or an extra copy of the der(18)t(14;18).

Follicular, predominantly large cell, intermediate grade lymphoma is also associated with the t(14;18)(q32;q24), which occurs in 56% of cases. Its presence is not predictive of survival. Gain of chromosome 7 occurs more frequently in large cell FL.

Mantle cell lymphoma (MCL): this rare type is refractory to treatment, and so having the diagnosis confirmed by cytogenetics or FISH is clinically helpful. A translocation t(11;14)(q13;q32), involving the *BCL1* gene at 11q13, is strongly associated with MCL *(8)*, and a genetic study for this translocation is often requested to confirm the diagnosis. It has also been described in other NHLs, such as lymphocytic lymphoma or centrocytic lymphoma, but these probably derive from (or have evolved from) MCL. It is not exclusive to the lymphomas, being one of the more common abnormalities in B-CLL and in multiple myeloma *(9)*. FISH detection of the t(11;14) has been reported to be more efficient than either conventional cytogenetics or molecular methods *(10)*. Other recurrent abnormalities in MCL include del(13)(q14), del(17)(p), and trisomy for all or part of chromosome 12. Within MCL, clone complexity and trisomy 12 appear to be indicators of a poorer prognosis *(8)*. Less common abnormalities in MCL include del(11)(q22–23) and t(11;22)(q13;q11).

Mucosa-associated lymphoid tissue lymphomas (MALT) and monocytoid B-cell lymphoma: These are low-grade lymphomas that occur typically in gastrointestinal tract, thyroid, breast, and skin. A recurrent but not exclusive abnormality is simple trisomy 7.

Splenic lymphoma with villous lymphocytes (SLVL): SLVL is a relatively benign disease but one in which a high incidence of clonal abnormalities has been reported; this is in contrast to most lymphoid malignancies, in which the detection of clones tends to be correlated with aggressive disease. Abnormalities of 14q32 are most common, and particularly a t(11;14)(q13;q32). There is some evidence that the gene on 11q13 is not the same as that involved in mantle cell lymphoma ***(11)***. There seems to be a close association between SLVL and abnormalities of 7q22 and/or 7q32, which are unusual in most other chronic lymphoid disorders (12,13). Trisomy 3 is also relatively common in SLVL.

Small lymphocytic, diffuse lymphoma: This kind frequently develops into leukemia. A common cytogenetic abnormality is deletion of part of 11q, especially involving band q23. This breakpoint is frequently involved in acute leukemias and is associated with an abnormality of the *MLL* gene; however, in lymphomas the significant gene appears to be *NCAM*, which is just proximal to *MLL* (*see also* **ref. 14**).

Small lymphocytic, plasmacytoid type lymphoma (LPL): This usually has an indolent course. It is associated with a t(9;14)(p13;q32), and involves the *PAX5* gene at 9p13 ***(15)***.

T-cell lymphomas: These are generally more common in children than in adults. As in T-ALL, nearly 50% of translocations involve the four known T-cell receptor sites: TCR-α and TCR-δ at 14q11, TCR-β at 7q34–36, and TCR-γ at 7p15. Translocations involving 9q34 are associated with intermediate to high-grade T-cell lymphomas. Other abnormalities that tend to be associated with T-cell lymphomas involve 1p, especially 1p22 and 1p36, and 6p21–24.

T-cell lymphoma of angioimmunoblastic lymphadenopathy type has also been associated with gain of an X chromosome, abnormalities of 1p31–32, and trisomy 3 (*see also* **ref. 16**).

Hodgkin's disease (HD): This is the most common kind of lymphoma, occurring particularly in young male patients. There are no specific chromosome abnormalities associated with HD, with the possible exception of abnormalities involving 4q25–27 ***(17)***. In general, cytogenetic studies of HD have a lower success rate and a lower clone detection rate. Some of the abnormalities that occur in NHL are also found in HD, 30% of cases having abnormalities of chromosome 1, 20% having translocations to 14q32, and 15% having del(6)(q). More common in HD than in NHLs are abnormalities involving 3q26–29, 7q, and 12p, and clones having near-triploid or near-tetraploid chromosome counts.

Childhood lymphomas: There are rather few published data for childhood lymphomas. These seem to be biologically and clinically different, the evidence being the rarity of FL, the greater proportion of high-grade lymphomas, and the increased incidence of mediastinal involvement. Consequently, a translocation t(9;17)(q34;q23), which is associated with mediastinal disease, is found almost exclusively in children and is very rare in adults ***(18,19)***.

3. Chronic Lymphoproliferative Disorders

The most common types of lymphoproliferative disorders (LPD) are chronic lymphocytic leukemia (CLL), prolymphocytic leukemia (PLL), and hairy cell leukemia (HCL). SLVL (*see* **Subheading 2.1.**) is also sometimes included in the LPD group.

CLL usually occurs late in life (median age 62) and can have a prolonged course (lasting many years) with little adverse effect on the patient; in many cases it is unsuspected and is discovered only during investigation for some other health problem. Aggressive treatment is generally contraindicated unless there are signs of atypical or advanced disease, as it may accelerate the progression of disease. Adverse signs include abnormalities involving 14q32, 17p, and 11q, and the presence of trisomy 12. Trisomy 12, which occurs in about 20% of cases, is probably not a primary abnormality associated with leukemogenesis of CLL, as it can be shown to be present in only a proportion of the CLL cells. It is associated with atypical CLL, with mixed CLL/PLL, or with disease progression. It is rarely seen in typical CLL. Because of its link with a poor prognosis, especially when it is not the sole abnormality, it has been inferred to be an indicator for active treatment ***(20)***. Del(11)(q21–25) has been found in up to 20% of cases by FISH; it is associated with a poor prognosis, and tends to occur in younger patients with extensive disease. Abnormalities of chromosome 13, usually involving band q14, are also common, occurring in about 10% of cases by cytogenetics and 50% of cases by FISH, and tend to be associated with typical CLL and therefore with a relatively good prognosis. The abnormalities of 13q are often subtle and may need to be detected by FISH rather than by conventional cytogenetics. Very few cases have been reported to have both trisomy 12 and abnormal 13q14, and generally these two abnormalities seem to be mutually exclusive. Studies of 13q14 are of current scientific interest since it is more likely to be a primary abnormality in the aetiology of CLL.

A small but important group of abnormalities are those that involve chromosome 17p; these have been shown to indicate a very poor prognosis ***(21,22)***.

As has been mentioned earlier, conventional cytogenetic studies of CLLs are often disappointing because there are frequently no divisions, and in cases where divisions are found these tend to be apparently normal. Consequently, there is a tendency to limit routine genetic analysis to the detection by FISH and reverse transcription-polymerase chain reaction (RT-PCR) of certain well established abnormalities such as trisomy 12, del(13)(q14), del (11)(q), and loss of P53 on 17p1. However, other recurrent abnormalities are still being discovered whose clinical significance has still to be determined; if the laboratory has the resources, a conventional cytogenetic study therefore can be worthwhile.

PLL is also usually a disease of older people, many >70 years of age at diagnosis. It can occur as a development of CLL. B-cell PLL is more common, although T-PLL forms a large proportion of the rare T-cell leukemias. Unlike CLL, PLL is a progressive disease with a poor prognosis; this is especially true for T-PLL, which is associated with a median survival of only 7.5 mo ***(23)***. Perhaps because of the advanced nature of the disease, cytogenetic studies in PLL have a much higher success rate, with a clone being found in almost all cases. In B-PLL, as in other B-cell malignancies, most clones have some abnormality of 14q32. Similarly, in T-PLL as in other T-cell disorders, there is frequent involvement of band 14q11.

However, an abnormality common in T-PLL but unusual elsewhere is an isochromosome for the long arms of chromosome 8, i(8)(q10) or idic(8)(p11) ***(24)***.

HCL is a B-lineage malignancy occurring predominantly in middle-age males. It is associated with widely-varying response to treatment, with survivals ranging from <2 yr to >10 yr. Rather few cytogenetic studies have been reported, and no characteristic abnormality has been described. Translocations involving 14q32 are common, as in other B-lineage malignancies.

4. Malignant Myeloma and Related Conditions

Malignant myeloma (MM) (also called multiple myeloma and myelomatosis) is a disease of B-lineage cells only; there is no T-cell myeloma. However, abnormal T-cell populations do occur in MM and are implicated in some of the clinical features, such as hypogammaglobulinemia and a susceptibility to infections. Although MM is described as a mature B-cell disease, the malignant cells are actually immature plasma cells, resembling plasmablasts. It is rare for B lymphocytes to be involved ***(25)***. The plasma cells secrete abnormally high levels of monoclonal immunoglobulin, which in some cases can be detected in the urine and is known as Bence-Jones protein. The cells may also form solid tumors, each a mass of plasma cells, that can occur anywhere in the body but particularly near the spine, where they can cause spinal cord compression.

MM is closely related to plasmacytoma and to plasma cell leukemia (PCL), a disease with a very poor prognosis, and aggressive forms of MM may evolve into PCL. Related but less severe are the immunoproliferative disorders, such as Waldenstrom's macroglobulinemia, and some patients with monoclonal gammopathy of unknown significance (MGUS) will later develop MM. As with the refractory anemias, the severity of these conditions correlates with the incidence of cytogenetically detected clones; in one series, the incidence of clones was 15% in WM, 25% in MGUS, 33% in MM, and 50% in PCL ***(26)***. However, many other factors influence the clone detection rate in MM. These include the percentage of plasma cells present (it is rare to find clones when there are fewer than 16% plasma cells in the bone marrow), the type of cell morphology (clones are rarer in small cell and Marschalko types, and much more common in cleaved, asynchronous, and blastic types ***[27]***), the clinical grade (18% in low grade up to 71% in high grade) ***(27–29)***, and the state (47% at diagnosis up to 71% in relapse, ***30***). Although cytogenetic results are of independent prognostic value ***(31,32)***, success in simply detecting any clone tends to be associated with poor-risk clinical features ***(26)***.

The MM cells divide slowly in comparison to the other cells present, and this is the main reason why no clone is detected by conventional cytogenetics in so many cases of MM. There have been various ways described that claim to improve the success rate, mainly by using cytokines such as B-cell mitogens and/or interleukins, and reported studies have varied in their conclusions. For these stimulated cultures, some have advocated 6-d incubation ***(30)***, and some as little as 2 d ***(33)***. Others have reported that unstimulated cultures have been better ***(29,34)***. The diversity of techniques is a consequence of their unpredictability; there appears to be no technique that is consistently reliable in all laboratories.

In most cases the clone found has been complex, which is taken to indicate that the disease is at an advanced stage by the time of diagnosis ***(35)***. The most common abnormalities found at diagnosis are loss of all or part of a chromosome 13, and translocations involving 1q (never as the sole abnormality), 8q24 (*c-MYC* gene), 11q13 (*BCL-1/cyclin D1*), 14q32 (*IgH*), 16q22 and 22q11 *IgL* ***(9)***. FISH (including spectral karyotyping [SKY] and multiplex FISH [M-FISH]) methods have revealed several recurring abnormalities involving 14q32 that are too subtle for detection by conventional cytogenetics ***(36)***; these include t(14;22)(q32;q11), t(14;16)(q32;q23), t(9;14)(p13;q32), and t(4;14)(p16;q32). Loss of the short arms of one chromosome 8 is also recurrent, whether by monosomy 8, del(8)(p) or by unbalanced translocations with a breakpoint at the centromere ***(36)***.

Deletions in 13q14 should be routinely sought using FISH or molecular methods, as they are usually too small to be seen on a conventional cytogenetic study. Identifying these is currently thought to be the most important result for the clinician, as they are the most significant abnormality in MM known so far that correlates with prognosis: patients with –13 or del(13)(q14) have a particularly poor prognosis ***(31,37)*** .

Although PCL and MM are closely related diseases, both having much the same type of cytogenetic abnormalities, there are some differences that suggest that they may not simply be different stages of the same disease. Both MM and PCL commonly have gain of the long arms of chromosome 1 and loss of all or part of chromosome 13, but these are more common in PCL. Clones are complex in both disorders, but are generally hyperdiploid in MM and hypodiploid in PCL. Loss of part of the short arms of chromosome 6 is rare in MM, but appears to be recurrent in PCL ***(38)***.

Because of the intensity of the treatment required to control MM, survivors are at risk of developing secondary myelodysplastic syndrome (MDS) and acute myeloid leukemia (AML). Therefore, if a patient has been undergoing treatment for MM for 2 yr or longer and has abnormal bone marrow cytology, a full conventional cytogenetic study should be performed. The type of chromosome abnormalities present may provide a differential diagnosis: for example, as mentioned, in MM the most common structural abnormalities involve 13q14 and 14q32, which are rarely involved in MDS/AML, while in secondary MDS/AML abnormalities of chromosomes 5 and 7 are most common. Abnormalities of chromosome 1 occur in both diseases, so the presence of these would not help to differentiate between the two diagnoses. In the author's laboratory, a patient was found to have two different, concurrent, complex clones, one with abnormalities consistent with MM, and the other having abnormalities typical of secondary AML. In sequential studies, the clones alternated in predominance.

5. Summary

The cytogenetic abnormalities that are found in chronic lymphoid malignancies (and in acute leukemias deriving from relatively mature cells) fall mainly into two groups according to whether the malignant cells are of B-lineage or T-lineage. In most of the B-lineage cases, there is some abnormality of the *IgH* gene which is located at 14q32 or, less frequently, the other immunoglobulin genes located at 2p12 and 22q11. In the T-lineage cases, there is often some abnormality involving the T-cell receptor loci, most frequently those at 14q11.

Other abnormalities occur, but few have a close association with a particular disease type, and so do not often contribute to determining a precise diagnosis. This lack of specificity is due partly to deficiencies in our understanding of the biological relationships between different lymphoid disorders, and also to the various classifications that have been used. As a consequence, there has to be some doubt about the diagnosis assigned to many of the cases in published cytogenetic studies. It is also difficult to combine data from series that have used different classification systems. The present unsatisfactory situation greatly limits the clinical usefulness of cytogenetic studies and there is a real need to unravel the complexities of this large family of malignancies.

References

1. Weisenburger, D. D., Gordon, B. G., Vose, J. M., et al. (1996) Occurrence of the t(2;5)(p23;q35) in non-Hodgkin's lymphoma. *Blood* **87,** 3860–3868.
2. Jaffe, E. S. (2001) Anaplastic large cell lymphoma: the shifting sands of diagnostic hematopathology. *Mod. Pathol.* **14,** 219–228.
3. Dyomin, V. G., Rao, P. H., Dalla-Favera, R., and Chaganti, R. S. (1997) *BCL8*, a novel gene involved in translocations affecting band 15q11–13 in diffuse large-cell lymphoma. *Proc. Natl. Acad. Sci. USA* **94,** 5728–5732.
4. Gilles, F., Goy, A., Remache, Y., Shue, P., and Zelenetz, A. D. (2000) *MUC1* dysregulation as the consequence of a t(1;14)(q21;q32) translocation in an extranodal lymphoma. *Blood* **95,** 2930–2936.
5. Juneja, S., Lukeis, R., Tan, L., et al. (1990) Cytogenetic analysis of 147 cases of non-Hodgkin's lymphoma: nonrandom chromosomal abnormalities and histological correlations. *Br. J. Haematol.* **76,** 231–237.
6. Fukuhara, S., Rowley, J. D., Variakojis, D., and Golomb, H. M. (1979) Chromosome abnormalities in poorly differentiated lymphocytic lymphoma. *Cancer Res.* **39,** 3119–3128.

7. Horsman, D. E., Connors, J. M., Pantzar, T., and Gascoyne, R. D. (2001) Analysis of secondary chromosomal alterations in 165 cases of follicular lymphoma with t(14;18). *Genes Chromosomes Cancer* 30, 375–382.
8. Cuneo, A., Bigoni, R., Rigolin, G. M., et al. (1999). Cytogenetic profile of lymphoma of follicle mantle lineage: correlation with clinicobiologic features. *Blood* **93,** 1372–1380.
9. Sawyer, J. R., Waldron, J. A., Jagannath, S., and Barlogie, B. (1995) Cytogenetic findings in 200 patients with multiple myeloma. *Cancer Genet. Cytogenet.* **82,** 41–49.
10. Li, J-Y., Gaillard, F., Moreau, A., et al. (1999) Detection of translocation t(11;14)(q13;q32) in mantle cell lymphoma by fluorescence in situ hybridization. *Am. J. Pathol.* **154,** 449–1452.
11. Troussard, X., Mauvieux, L., Radford-Weiss, I., et al. (1998) Genetic analysis of splenic lymphoma with villous lymphocytes: A Groupe Francais d'Hematologie Cellulaire (GFHC) study. *Br. J. Haematol.* **101,** 712–721
12. Oscier, D. G., Matutes, E., Gardiner, A., et al. (1993) Cytogenetic studies in splenic lymphoma with villous lymphocytes. *Br. J. Haematol.* **85,** 487–491.
13. Oscier, D. G., Gardiner, A., and Mould, S. (1996) Structural abnormalities of chromosome 7q in chronic lymphoproliferative disorders. *Cancer Genet. Cytogenet.* **92,** 24–27
14. Caraway, N. P., Du, Y., Zhang, H-Z., Hayes, K., Glassman, A. B., and Katz, R. L. (2000) Numeric chromosomal abnormalities in small lymphocytic and transformed large cell lymphomas detected by fluorescence in situ hybridization of fine-needle aspiration biopsies. *Cancer* **90,** 126–132.
15. Offit, K., Parsa, N. Z., Filippa, D., Jhanwar, S. C., and Chaganti, R. S. K. (1992) t(9;14)(p13;q32) denotes a subset of low-grade non-Hodgkin's lymphoma with plasmacytoid differentiation. *Blood* **80,** 2594–2599.
16. Schlegelberger, B., Zwingers, T., Hohenadel, K., et al. (1996) Significance of cytogenetic findings for the clinical outcome in patients with T-cell lymphoma of angioimmunoblastic lymphadenopathy type. *J. Clin. Oncol.* **14,** 593–599.
17. Atkin, N. B. (1998) Cytogenetics of Hodgkin's disease. *Cytogenet. Cell Genet.* **80,** 23–27.
18. Kaneko, Y., Frizzera, G., Maseki, N., Sakurai, M., Komada, Y., et al. (1988) A novel translocation, t(9;17)(q34;q23), in aggressive childhood lymphoblastic lymphoma. *Leukemia* **2,** 745–748.

19. Shikano, T., Arioka, H., Kobayashi, R., et al. (1994) Acute lymphoblastic leukemia and non-Hodgkin's lymphoma with mediastinal mass-a study of 23 children; different disorders or different stages? *Leukemia Lymphoma* **13,** 161–167.
20. Juliusson, G., Oscier, D. G., Fitchett, M., et al. (1990) Prognostic subgroups in B-cell chronic lymphocytic leukemia defined by specific chromosomal abnormalities. *N. Eng. J. Med.* **323,** 720–724.
21. Lens, D., De Schouwer, P. J. J. C., Hamoudi, R. A., et al. (1997) p53 abnormalities in B-cell prolymphocytic leukemia. *Blood* **89,** 2015–2023.
22. Dohner, H., Stilgenbauer, S., Benner, A., et al. (2000) Genomic aberrations and survival in chronic lymphocytic leukemia. *N. Engl. J. Med.* **343,** 1910–1916.
23. Dearden, C. E., Matutes, E., Cazin, B., et al. (2001) High remission rate in T-cell prolymphocytic leukemia with CAMPATH-1H. *Blood* **98,** 1721–1726.
24. Sorour, A., Brito-Babapulle, V., Smedley, D., Yuille, M., and Catovsky, D. (2000) Unusual breakpoint distribution of 8p abnormalities in T-prolymphocytic leukemia: a study with YACS mapping to 8p11–p12. *Cancer Genet. Cytogenet.* **121,** 128–132.
25. Zandecki, M., Bernardi, F., Genevieve, F., et al. (1997) Involvement of peripheral blood cells in multiple myeloma: chromosome changes are the rule within circulating plasma cells but not within B lymphocytes. *Leukemia* **11,** 1034–1039.
26. Calasanz, M. J., Cigudosa, J. C., Odero, M. D., et al. (1997) Cytogenetic analysis of 280 patients with multiple myeloma and related disorders: primary breakpoints and clinical correlations. *Genes Chromosomes Cancer* **18,** 84–93.
27. Weh, H. J., Bartl, R., Seeger, D., Selbach, J., Kuse, R., and Hossfeld, D. K. (1995) Correlations between karyotype and cytologic findings in multiple myeloma. *Leukemia* **9,** 2119–2122.
28. Dewald, G. W., Kyle, R., Hicks, G. A., and Griepp, P. R. (1985) The clinical significance of cytogenetic studies in 100 patients with multiple myeloma, plasma cell leukemia, or amyloidosis. *Blood* **66,** 380–390.
29. Smadja, N. V., Louvet, C., Isnard, F., et al. (1995) Cytogenetic study of multiple myeloma at diagnosis: comparison of two techniques. *Br. J. Haematol.* **90,** 619–624.
30. Luc Lai, J., Zandecki, M., Mary, J. Y., et al. (1995) Improved cytogenetics in multiple myeloma: A study of 151 patients including 117 patients at diagnosis. *Blood* **85,** 2490–2497.

31. Tricot, G., Barlogie, B., Jagannath, S., et al. (1995) Poor prognosis in multiple myeloma is associated only with partial or complete deletions of chromosome 13 or abnormalities involving 11q and not with other karyotype abnormalities. *Blood* **86,** 4250–4256.
32. Tricot, G., Sawyer, J. R., Jagannath, S., et al. (1997) Unique role of cytogenetics in the prognosis of patients with myeloma receiving high-dose therapy and autotransplants. *J. Clin. Oncol.* **15,** 2659–2666.
33. Cuneo, A., Balsamo, R., Roberti, M. G., et al. (1996) Interleukin-3 plus interleukin-6 may improve chromosomal analysis of multiple myeloma: Cytologic and cytogenetic evidence in thirty-four patients. *Cancer Genet. Cytogenet.* **90,** 171–175.
34. Brigaudeau, C., Trimoreau, F., Gachard, N., et al. (1997) Cytogenetic study of 30 patients with multiple myeloma: comparison of 3 and 6 day bone marrow cultures stimulated or not by using a miniaturized karyotypic method. *Br. J. Haematol.* **96,** 594–600.
35. Seong, C., Delasalle, K., Hayes, K., et al. (1998) Prognostic value of cytogenetics in multiple myeloma. *Br. J. Haematol.* **101,** 189–194.
36. Sawyer, J. R., Lukacs, J. L., Thomas, E. L., et al. (2001) Multicolour spectral karyotyping identifies new translocations and a recurring pathway for chromosome loss in multiple myeloma. *Br. J. Haematol.* **112,** 167–174.
37. Desikan, R., Barlogie, B., Sawyer, J., et al. (2000) Results of high-dose therapy for 1000 patients with multiple myeloma: durable complete remissions and superior survival in the absence of chromosome 13 abnormalities. *Blood* **95,** 4008–4010.
38. Gutierrez, N. C., Hernandez, J. M., Garcia, J. L., et al. (2001) Differences in genetic changes between multiple myeloma and plasma cell leukemia demonstrated by comparative genomic hybridization. *Leukemia* **15,** 840–845.

9

Other Lymphoid Malignancies

Cytogenetic Techniques

John Swansbury

1. Introduction

This chapter describes the practical aspects of performing cytogenetic studies in a variety of lymphoid disorders, including the lymphomas, multiple myeloma, chronic lymphocytic leukemia, and other chronic lymphoproliferative diseases. They are also required for studies of acute lymphoblastic leukemia of mature T-cell or B-cell types. As mentioned in the previous chapter, most lymphoid cells are either T-lineage or B-lineage. During normal differentiation, they become capable of responding to antigens, and one of these responses is to transform and undergo division. A variety of reagents (known as mitogens) with antigenic properties are used in the laboratory to stimulate the cells to transform in a similar way. The two mitogens featured in the methods described here are Phytohemagglutinin (PHA) for T-cells (*see* **Note 1**), and 12-*O*-tetradecanoyl-phorbol-13-acetate (TPA) for B cells. Other B-cell mitogens are described in **Note 2**. Be aware that no mitogen is absolutely specific to T cells or B cells (*see* **Note 3**).

From: *Methods in Molecular Biology, vol. 220: Cancer Cytogenetics: Methods and Protocols*
Edited by: John Swansbury © Humana Press Inc., Totowa, NJ

2. Materials

This list is very similar to that described in Chapter 4, and more details are given in that chapter. *Note: Many of the reagents and chemicals used can be harmful and should be handled with due care and attention.* Always refer to the information provided by the manufacturer. Most of the solutions should be kept in the dark at about 4°C. The dilutions given here of most of the reagents are such that 0.1 mL (100 μL) may be conveniently added to a 10-mL culture. Except for PHA, the solutions should be filter sterilized (e.g., with a 0.22-μm millipore filter).

1. Containers: sterile, capped, plastic, 10-mL centrifuge tubes.
2. Pipets: plastic, disposable.
3. Medium: RPMI 1640 (GIBCO) with Glutamax is recommended. Many other media may be used successfully, such as TC199, McCoy's 5A, and Ham's F10, but RPMI was developed specifically for leukemic cells. To each 100-mL bottle, add antibiotics (e.g., 1 mL of penicillin + streptomycin) and 1 mL of preservative-free heparin. If the medium does not contain Glutamax then L-glutamine should be added (final concentration 0.15 mg/mL).
4. Serum: fetal calf serum; the proportion routinely added is 15 mL of serum to 100 mL of medium.
5. Mitogens: care should be taken to ensure that these solutions do not become contaminated. They should be stored in small quantities (2 mL) at –20°C until needed.
   a. PHA: obtained freeze-dried or lyophilized, ready for reconstitution to the appropriate concentration with sterile water.
   b. TPA : obtained as a powder. The solution is prepared by dissolving in 10% ethanol and then further diluting 1:19 with water to make a stock solution of 10 μg/mL in 0.5% ethanol. This must be stored frozen in small quantities. It is light-sensitive, so the containers should be securely wrapped in foil.
6. Blocking agent: excess thymidine (XT) ***(1)*** is preferred for lymphoid disorders, although fluorodeoxyuridine (FdUr) (recommended for myeloid disorders and described in Chapter 4) is also often successful. Dissolve 1 g of thymidine (Sigma) in 33 mL of buffer to make a 0.05 m*M* stock solution. Store frozen in 2-mL volumes. Once thawed, do not refreeze.

7. Releasing agent: deoxycytidine: Dissolve 10 mg in 22 mL of buffer to make a 10 µ*M* stock solution.
8. Arresting agent: colcemid (also called demecolchicine, from deacetyl-methylcolchicine): 10 µg/mL stock solution.
9. Hypotonic solution: 0.075 *M* Potassium chloride (KCl, 5.59 g/L). Use at 37°C.
10. Fixative: three parts absolute methanol and one part glacial acetic acid. This should be freshly prepared just before use although it may be kept for a few hours if chilled.
11. 2.5% Trypsin (Dulbecco): dtored frozen in 1-mL aliquots. Diluted 1:50 in buffer (Ca^{2+}- and Mg^{2+}-free, for example, Hanks' buffered salt solution) when required.
12. Phosphate-buffered saline: pH 6.8, used for diluting stain.
13. Slides: the frosted-end variety are preferable for convenience of labeling. The slides must be free of dust and grease. Specially cleaned slides may be bought, otherwise wash in detergent, rinse well in water, then in dilute hydrochloric acid and alcohol.
14. Stains: Wright's stain. This is diluted immediately before use 1:4 with pH 6.8 buffer. (Giemsa and Leishman's stains are also suitable.)
15. Coverslips: 22 × 50 mm, grade 0 preferred.
16. Mounting medium: Gurr's neutral mounting medium is routinely used in this laboratory.
17. Incubator: the need for a CO_2-controlled, humidified incubator at 37°C is greater for lymphoid disorders than for myeloid disorders, as cultures with mitogens need several days for the cells to respond. The caps of the culture tubes must be loosened to allow diffusion of gas. However, good results are usually obtained with a simple incubator that has only temperature control if the cultures are gassed with 4% CO_2 in air, which also helps to maintain the pH of the culture medium if it is bicarbonate buffered.
18. Laminar flow cabinet: because mitogen-stimulated cultures are grown for several days, there is a greater chance of infection taking hold. Therefore sterile technique is more important.

3. Methods

3.1. Types of Sample

Wherever possible, the tissue sent for cytogenetic study should be from the tumor, rather than from some other tissue that may be infiltrated with lymphoid cells. In particular, a conventional cytogenetic study cannot be used to demonstrate that a secondary tissue is *not* infiltrated.

Lymph node biopsy is the preferred tissue for studies of most lymphomas. It should be transported in culture medium and it needs prompt attention: The failure rate is very high unless cultures are set up on the same day that the node was removed. On receipt in the laboratory, the node should be washed in medium containing five times the usual concentration of antibiotics. Using full sterile technique, place it in a Petri dish in a small volume of fresh medium, and remove any extraneous material (such as fat and connective tissue) or necrotic material (usually at the center of the tumor). Using sterile scissors or a scalpel, cut through the biopsy specimen. Some nodes release large numbers of white cells freely, and further cutting into small pieces will release enough cells for culture. Other nodes tend to be very tough, sclerous (rubbery), and difficult to handle. Some cells will be obtained by mincing the sample into small pieces; another approach is to force the pieces through a sterile, fine wire gauze.

Lymph node aspirates are sometime sent for cytogenetic study. These rarely provide enough cells for the variety of cultures that can be set up from a lymph node biopsy; if there are only enough cells for one culture, then 4-h or overnight colcemid is best. If the aspirate contains very few cells (e.g., less than enough for even one culture) then the chances of success are very small.

A *spleen* sample for cytogenetic study may be sent after splenectomy, a clinical intervention used in some malignancies partly to reduce the tumor load and alleviate disease side effects. This is treated in the same way as a lymph node to release cells for culture.

A *bone marrow* aspirate or trephine can be successful for investigation of lymphoid disorders if it is sufficiently infiltrated with malignant cells. In the author's laboratory, clonal cells have been found on rare occasions even in the absence of cytological evidence of infiltration. However, this occasional success does not justify routine studies on bone marrow that is not obviously infiltrated.

Blood samples have a higher failure rate (especially for unstimulated cultures), and a lower clone detection rate, but are frequently sent for investigation of lymphoid disorders. If the white cell count is low, the sample should be centrifuged and the cell-rich top layer used; having too many red blood cells can interfere with processing during harvesting.

3.2. Setting Up Cultures

Careful sterile technique should be developed, as many cultures need to be maintained for 3–5 d, which is ample time for contaminating organisms to proliferate even in the presence of antibiotics.

1. Perform a cell count, if possible, and use 8–10-mL cultures with a cell density of about 1×10^6/mL. Be aware that an automated cell counter may overestimate the number of viable cells present; a manual count using Trypan blue staining can give a more reliable estimate.
2. The cytogenetic study should include a variety of cultures, both unstimulated and mitogen-stimulated. It is recognized that many laboratories do not have the time and resources to set up, process, and screen all the cultures that might work! For example, seven to nine cultures are described below for B-cell disorders. Each laboratory therefore should test these on a few samples and assess what works best.

   Set up the following cultures (summarized in **Table 1**):

   a. For all patients: Four unstimulated cultures, one with overnight colcemid, one blocked overnight with excess thymidine, one cultured overnight before colcemid is added the next day for harvesting an hour later, and one for harvesting after 3-5 d of incubation. This long-term, unstimulated culture has been claimed to be particularly useful for selectively obtaining divisions from the malignant plasma cells in multiple myeloma *(2)* as it is unfavorable for normal cells and therefore increases the likelihood of the divisions being from plasma cells.
   b. For all patients: Two cultures with PHA, to be harvested after 3 d, to be used to check the patient's constitutional karyotype if necessary (*see* **Notes 4** and **5**).

Table 1
Recommended Cultures for T- and B-Lymphoid Disorders

T-cell disorders			B-cell disorders			
Likely diagnoses: Adult T-cell leukemia/lymphoma (ATLL) T-cute lymphocytic leukemias T-non-Hodgkin lymphoma Sezary syndrome Mycosis fungoides T-prolymphocytic leukemia (T-PLL)			*Likely diagnoses*: Most lymphomas Most chronic lymphocytic leukemias L3 type acute lymphocytic leukemias Hairy cell leukemia Myeloma Plasma cell leukemia Waldenstrom's macroglobulinemia Monoclonal gammopathy of unknown significance (MGUS)			
(a) Unstimulated			(a) Unstimulated			
Colcemid overnight	Excess thymidine block overnight	Overnight culture, then add colcemid for 1–2 h.	Colcemid overnight	Excess thymidine block overnight	Overnight culture, then add colcemid for 1–2 h.	4–5-d culture, then add colcemid for 1–2 h.

(b) PHA-stimulated			(b) TPA (or other B-mitogen)-stimulated		
Culture for 3–4 d, then add colcemid overnight; harvest next morning.	Culture for 3–4 d, then excess thymidine block overnight.	Culture for 3–4 d, then add colcemid for 1 h before harvesting.	Culture for 4–5 d, then add a reduced volume of colcemid overnight; harvest next morning.	Culture for 4–5 d, then excess thymidine block overnight; use a reduced volume of colcemid to collect divisions.	Culture for 4–5 d, then add colcemid for 1 h before harvesting.

Other cultures should be set up if there is enough material. After adding colcemid and other reagents, mix well by inverting the tube. For stimulated cultures, enter in the laboratory diary the date when the overnight reagents should be added.

c. For patients with T-cell disease: Two or three more cultures with PHA, to be harvested after 4–5 d. In some cases the malignant T cells have a poor immunologic response, and they may react better to a cocktail of mitogens, such as PHA + pokeweed mitogen (PWM) + TPA.
d. For patients with B-cell disease: Three or more cultures with TPA. If there is enough spare material available, set up a further set of cultures either with another B-cell mitogen or else with TPA but harvested on a different day. B cells need to be cultured for at least 3 d for mitogens to have an effect; it can be helpful to have duplicate cultures grown for up to 5 and 7 d. In the author's laboratory, the usual practice is to have mitogen-stimulated 4-d cultures, and preferably further cultures grown for 1 or 2 d longer.

   After the incubation, the three cultures will be harvested after short exposure to colcemid, overnight colcemid, and after blocking with excess thymidine, respectively. Therefore, label each tube with the date, the patient's ID, the mitogen, the time in culture, and the type of harvesting. Add the measured volume of sample with the appropriate number of cells, and then add the complete medium to a volume of 8-10 mL. Add the mitogen(s) as required and mix thoroughly by inverting the tubes several times. It is worth noting in a laboratory diary when each type of culture is due to be harvested, so that it will not be forgotten. If it is not the laboratory policy to harvest cultures at weekends, but culture times require such harvests, then either put the cultures overnight in the refrigerator before putting them into the incubator, as this will delay the start of the culture time and is not usually detrimental to the cells; or else let the cultures continue for longer; 1 or 2 extra days should be acceptable.

3. At the end of the afternoon, add 100 μL of colcemid to one of the unstimulated cultures, and 100 μL of the excess thymidine solution to the other.
4. Stand the mitogen-stimulated culture tubes in the incubator (at 37°C) at an angle, rather than upright, as this increases the surface area of the deposit and reduces local exhaustion of the medium. If there is very little air space in the top of the tube, there will be insufficient oxygen to last throughout the incubation period. Loosen the caps of the tubes to allow diffusion of gases. However, for best culture conditions, the amount of oxygen in the atmosphere is too high and the amount of CO_2 is too low. If the incubator is CO_2-controlled, then

this provides a better balance of gases for the cells in culture. If it is not, then *see* **Note 6**.

5. Next morning harvest the colcemid overnight culture, using the procedure described in Chapter 4. To the culture with excess thymidine, add 100 µL of the releasing agent (deoxycytidine) (*see* **Note 7**) and 100 µL of colcemid (*see* **Note 8**), mix, and then return to the incubator for 4 h. To the culture that was incubated overnight without additives, add colcemid for 1–4 h (*see* **Note 9**). At the end of this time, harvest in the same way as the overnight colcemid culture.
6. For the mitogen-stimulated cultures: On the evening before the end of the required number of days in the incubator, add colcemid to one of each trio of cultures, and excess thymidine to another. The next day, add colcemid to the third tube, for harvesting after a short exposure time. The harvesting and processing are as for the unstimulated cultures. Chromosomes of divisions from TPA-stimulated cultures appear to be extra-sensitive to colcemid, and it is worth using a shorter exposure time, or reduced volume of colcemid (e.g., 20–40 µL) for these cultures.

3.3. Harvesting

All the harvesting, processing, spreading, banding, and staining may be performed in the same way as described in Chapter 4.

4. Notes

1. T-cell mitogens: PHA (Phytohemagglutinin) is the most specific and widely used mitogen for T lymphocytes. It acts via monocytes which produce interleukin-2 (IL-2), and so some centers add IL-2 as well as PHA in case the monocyte response is affected by the patient's disease.

   PHA is widely regarded as being T-cell specific, but clonal cells in B-cell disorders can be found after PHA stimulation, due to a recognized response known as T-cell-dependent B-cell activation *(3)*.

   The action of PHA on *normal* T cells will usually produce many divisions within 48 h, and in 72-h cultures it is possible for some cells to have progressed to a second division. However, malignant cells will often have a slower response and 48 h may not be long enough. Similarly, there may be a very poor response to mitogens if the patient has started chemotherapy, which is immunosuppressive. If possible, leave the cultures until the color of the medium has become light yellow-amber and there is a dark edge to the red cell sediment.

Occasionally a sample may be sent in a tube containing EDTA instead of heparin. If possible, decline to accept this sample and ask for another sample that has been placed in heparin. If, however, a replacement cannot be obtained, then it is necessary to wash out the EDTA by going through two cycles of centrifugation and resuspending in fresh, complete culture medium. When setting up the cultures, increase the cell concentration by about 20% to compensate for the effect that EDTA will have had. EDTA inhibits the response to PHA, so no divisions will be obtained unless it is removed.

2. B-cell mitogens: The most commonly used mitogens are:
   a. TPA, also called phorbol 12-myristate 13-acetate (PMA).
   b. PWM
   c. Lipopolysaccharide (LPS), obtained from *E. coli.*
   d. Protein A
   e. IL-6 (can be used for myeloma cells).
   f. Epstein-Barr virus (EBV): This virus is a potential pathogen and should be grown only under carefully controlled conditions; some centers do not permit its use. If some of the supernatant from an EBV culture is available to the cytogenetics laboratory from a local virology unit, pass it through a 0.22-μm Millipore filter, and use to make 10% of the volume of the leukocyte culture.

   After many years of trying different B-cell mitogens, it has been found that TPA has given the highest clone detection rate in the author's laboratory. This may not be true for other laboratories. There has not been agreement between published studies about the most effective B-cell mitogens; for example, PWM gave a very poor result in some studies ***(4)*** but was one of the best mitogens in others ***(5)***; the converse was found with TPA. It may well be that success or failure depends on other, unidentified, or local factors as much as on the choice of mitogen ***(6)***.
3. Most B-cell mitogens will also stimulate T cells, so the divisions obtained from a B-mitogen-stimulated culture cannot be assumed to be exclusively from that lineage. Because the normal cells may have a greater response, and the malignant cells often have a low mitotic index, it may be necessary to analyze a relatively large number of divisions to improve the chances of detecting a clone. This is particularly true of a disease in its early stages. In a well advanced malig-

nancy, the clonal cell population tends to suppress the formation of normal cells, and then all the divisions found may be abnormal. Some laboratories prefer to use a "cocktail" of several B-cell mitogens in the same culture. In our experience, this can increase the likelihood of stimulating unwanted normal cells into division.

4. Determining a patient's constitutional karyotype: Most laboratories routinely set up PHA-stimulated cultures on all new cases to obtain plentiful, good quality metaphases that can be used to determine the patient's constitutional karyotype. This is important in case all the divisions obtained from other cultures have the same abnormality, especially if it is one that is not usually known to be associated with the patient's diagnosis (*see* **Note 3**). However, if the patient has a T-cell disease, then some or all of the divisions in the PHA-stimulated cultures may derive from a clone. Furthermore, diseases of hemopoietic stem cells or early pluripotent progenitor cells (e.g., chronic myeloid leukemia) may also result in abnormal divisions being found in PHA-stimulated cultures. Their presence does not mean that they are constitutional. If the constitutional karyotype cannot be reliably determined using a blood sample, then another tissue must be used. The most usual is a skin biopsy.
5. A case history: Cytogenetic studies in two laboratories (at the Mayday Hospital, Croydon, UK [Ms. Carol Brooker], and in the author's laboratory) of a patient with a diagnosis of acute myeloid leukemia (AML) found a karyotype 46,XX,t(14;18)(q32;q21) in all divisions. This abnormality is usually associated with follicular lymphoma, not AML, and so the possibility of a misdiagnosis or of multiple diagnoses had to be considered. The same abnormality was seen in divisions from stimulated cultures, and was subsequently found in studies made of samples taken in remission, implying that it was constitutional. There were no other family members to study. A skin biopsy was grown for 3 mo, and all the divisions had the same abnormality, confirming that it was indeed constitutional. Finally, molecular studies showed that the breakpoints of this t(14;18) were not the same as those in the t(14;18) found in lymphoma.
6. Gassing cultures: It is important that the pH of the culture medium remains neutral or cell division will be inhibited. The normal color range is from an orange-peach color through to a pale-yellow or straw color. A vivid yellow often indicates the presence of an infection. More commonly, the color of the culture medium tends to become pink, and this can be rectified by increasing the amount of CO_2. A

simple, if crude procedure is as follows: Connect a source of carbon dioxide (e.g., a compressed gas cylinder) with tubing to a closed flask containing a solution of copper sulfate such that the gas bubbles through the solution. From the flask run another length of tubing into which a sterile pipet or sterile plastic quill can be inserted. Adjust the volume of gas passing through so that there is a steady flow. Remove the cap of the culture tube and let the gas fill the space above the medium for a few minutes. Do not create bubbles, as these are likely to increase opportunities for infection. Replace the cap of the tube and return the tube to the incubator for 10–15 min before reassessing the color.

7. If deoxycytidine is unavailable, simply centrifuge the culture, remove the old medium, and replace it with warmed, fresh medium. This will wash out the excess thymidine.
8. The published protocols for using excess thymidine recommend adding deoxycytidine alone at the start of the release time, with colcemid being added for just the last 15–20 min before harvesting *(1)*. If this procedure is used, in our experience it is necessary to double the amount of colcemid added to 200 μL, otherwise the divisions are few and poorly spread.
9. Although it is not usual to get any divisions from unstimulated, short-term cultures of normal lymphoid cells, they do sometimes occur. This is usually because they had been previously stimulated into transformation by a cause unrelated to the malignant disease, such as an infection. Therefore, the finding of only normal cells has to be interpreted with caution.

References

1. Wheater R. F. and Roberts S. H. (1987) An improved lymphocyte culture technique: deoxycytidine release of a thymidine block and use of a constant humidity chamber for slide making. *J. Med. Genet.* **24,** 113–115.
2. Smadja N. V., Louvet C., Isnard F., et al. (1995) Cytogenetic study of multiple myeloma at diagnosis: comparison of two techniques. *Br.J.Haematol.* **90,** 619–624.
3. Sole, F., Woessner, S., Perez-Losada, A., et al. (1997). Cytogenetic studies in seventy-six cases of B-chronic lymphoproliferative disorders. *Cancer Genet. Cytogenet.* **93,** 160–166.

4. Gahrton G. and Robert K. H. (1982) Review article: chromosome aberrations in chronic B-cell lymphocytic leukemia. *Cancer Genet. Cytogenet.* **6,** 171–181.
5. Sadamori N., Matsui S. I., Han T., and Sandberg A. A. (1984) Comparative results with various polyclonal B-cell activators in aneuploid chronic lymphocytic leukemia. *Cancer Genet. Cytogenet.* **11,** 25–29.
6. Connor, T. W. E. (1985) Phorbol ester-induced loss of colchicine sensitivity in chronic lymphocytic leukemia lymphocytes. *Leukemia Res.* **9,** 885–895.

10

Cytogenetic and Genetic Studies in Solid Tumors

Background

John Swansbury

1. Introduction

Solid tumors comprise approx 95% of all malignancies, but account for only a little over 25% of cases in published cytogenetic studies. The main reasons are:

1. It is more difficult to obtain samples of malignant tissue, which often require an operation rather than taking a simple blood sample or bone marrow aspirate.
2. There are technical difficulties in obtaining dividing cells from the tissue, which sometimes need long-term tissue culture with all its susceptibility to the problems of infection and contamination.
3. The karyotypes are generally more complex, as most solid tumors are at an advanced stage by the time they are diagnosed; consequently it is unclear what abnormalities have occurred early, which are associated with disease progression, and which are a late, random result of increasing genetic instability.
4. The clinical usefulness of the results is limited, compared to that of studies in leukemias, so there has been less incentive to devote to

From: *Methods in Molecular Biology, vol. 220: Cancer Cytogenetics: Methods and Protocols*
Edited by: John Swansbury © Humana Press Inc., Totowa, NJ

solid tumors the scarce time and resources of a service-based cytogenetics laboratory.

The situation has improved over the last decade (*see* **Fig. 1** in Chapter 1), partly as a result of the increased clinical expectations arising from an awareness of the value of cytogenetic studies in leukemias, and partly because of the successful application of fluorescence *in situ* hybridization (FISH) and molecular techniques. These have permitted studies even of sections from paraffin-embedded blocks. As a consequence, the proportion of genetic and cytogenetic publications relating to solid tumors is increasing rapidly. Interphase FISH studies are better for following-up previously identified abnormalities in well studied types of tumor ***(1)*** but comparative genomic hybridization (CGH) is better for investigating new tumors or new patients ***(2)***. Despite their technical difficulties, conventional cytogenetic studies of solid tumors still provide information that cannot be obtained by FISH or CGH. If at all possible, they should always be done on new samples.

Because it is such a wide and varied subject, this chapter cannot give a comprehensive overview of the whole range of cytogenetic abnormalities in solid tumors. It will concentrate on some of those that have been shown to have particular diagnostic or prognostic usefulness, and on some of the associations discovered since the publication of the excellent book *Cancer Cytogenetics,* by Heim and Mitelman ***(3)***. Although this book is becoming increasingly dated, it has not yet been surpassed as a summary of what has been known.

2. Selected Types of Solid Tumors Associated With Known Cytogenetic Abnormalities

1. Breast cancer: chromosome 17 carries genes that are associated with inherited predisposition to breast cancer. These include the *BRCA1* gene at 17q12, and the *p53* gene at 17p13 which is deleted in families with the Li-Fraumeni syndrome. Acquired abnormalities of chromosome 17 are also present in karyotypes obtained from malignant cells, and amplification of the *HER-2/neu* oncogene has been reported to be associated with a poor prognosis ***(4)***.

2. Clear cell sarcoma: A recurrent abnormality is t(12;22)(q13;q12), with fusion of the *ATF1* and *EWS* genes.
3. Desmoplastic small round cell tumors: A consistent abnormality is t(11;22)(p13;q12), which involves the *WT1* and *EWS* genes.
4. Germ cell tumors: an isochromosome for the short arms of a chromosome 12, i(12)(p10), is closely associated with germ cell tumors, and these tend to respond well to treatment with agents such as germ cell tumors, and these tend to respond well o platinum-based treatments. Cytogenetic and FISH studies using probes for 12p and 12q have been shown to be helpful in distinguishing between germ cell and other tumors in patients with poorly differentiated carcinoma; consequently these patients benefited from receiving the appropriate treatment ***(5)*** and other patients can be spared unsuitable treatment. Loss of a chromosome 3 has been shown to indicate a poor prognosis ***(6)***.
5. Myxoid liposarcoma: Consistent abnormalities are t(12;16)(q13;p11) with fusion of the *CHOP* and *FUS* genes, and t(12;22)(q13;q12), with fusion of *CHOP* and *EWS*.
6. Prostatic cancer with a more aggressive behavior is significantly associated with changes involving chromosomes 7, 8, X, and Y, and detection of these has a major impact on therapeutic decisions ***(7)***.
7. Squamous cell carcinoma of the head and neck has one of the lowest 5-yr survival rates for solid tumors. Amplification of the *cyclin D1* gene at 11q13, detected by using FISH, appears to be associated with poorly differentiated tumors and metastasis ***(8)***.
8. Synovial sarcoma: A highly specific cytogenetic abnormality, t(X;18)(p11;q11), is found almost exclusively in this tumor. Molecular studies using reverse transcription-polymerase chain reaction (RT-PCR) have identified three types of this translocation, which correspond to different clinical outcomes ***(9)***.
9. Transitional cell tumors of the bladder. Few samples could be more easily available for study than urine, and FISH studies of cells collected from urine samples have been successful in identifying malignant tumors ***(10)***. This confirmation of diagnosis is particularly helpful in cases with benign conditions such as papilloma or the secondary effects of cystitis, which can be difficult to identify by cystoscopy or by the standard morphological examination of cell cytology.
10. Cytogenetic studies of brain tumors have already discovered some specific abnormalities, such as a der(1)t(1;22)(p11;q12) in meningiomas occurring after radiotherapy ***(11)***.

11. The small blue cell tumors of childhood: These are tumors that tend to occur in children and have a similar cell morphology, so it is difficult to make a firm differential diagnosis based on morphology alone. Included in this group are rhabdomyosarcoma, Ewing's sarcoma, mesenchymal chondrosarcoma, small-cell osteosarcoma, hemangiopericytoma, neuroblastoma, and the peripheral neurectodermal tumors. Determining the precise diagnosis might not be so important if all such tumors had the same prognosis and treatment. However, they do not. Fortunately several genetic and cytogenetic abnormalities have now been identified that are disease specific for some of these malignancies, or that can indicate a different prognosis ***(12)***. These are included in the following list.
    a. Alveolar rhabdomyosarcoma: this is closely associated with a consistent chromosomal translocation, t(2;13)(q35;q14). The genes involved have been identified and are *PAX3* at 2q35 and *FKHR* at 13q14. A variant is t(1;13)(p36;q14), involving the *PAX7* gene at 1p36.
    b. Embryonal rhabdomyosarcoma: no specific abnormality has been associated with this disease ***(13)***. However, there are preferential gains of chromosomes 2, 8, 12, and 13, and about one third of cases have rearrangements in the region 1p11–1q11.
    c. Ewing's sarcoma and peripheral primitive neuroectodermal tumors: a specific translocation has been found, t(11;22)(q24;q12), which results in the fusion of the *FLI1* and *EWS* genes. There are also variant translocations, all involving *EWS* with partner genes at 21q12, 7p22, 2q33, and 17q22. Well over 90% of Ewing's sarcoma samples studied have been found to have abnormality of the *EWS* gene at 22q12. A common secondary abnormality is trisomy 8, and an unbalanced translocation, der(16)t(1;16)(q21;q13), is also recurrent. The clinical significance of these is not known. The *EWS* gene is also involved in other translocations that are associated with different malignancies, including t(12;22)(q13;q12) in myxoid liposarcoma and clear-cell sarcoma, and t(11;22)(p13;q12) in desmoplastic small round cell tumor.
    d. Neuroblastoma: The most common abnormality is deletion of part of the short arms of a chromosome 1 in the area around 1p32–36, and this is usually associated with more progressive disease.
    e. Wilms' tumor: This is a tumor of the kidney and, like retinoblastoma, is most common in children who have an inherited genetic abnormality. Two genes are commonly involved, the *WT1* gene

at 11p13, which is associated with the WAGR syndrome, and the *WT2* gene at 11p15, which is associated with the Wiedemann–Beckwith syndrome. However, there is no common, specific cytogenetic abnormality associated with Wilms' tumor. The prognosis is generally good, with >80% survival. Loss of all or part of chromosome 22 is a recurrent abnormality that is associated with a poor prognosis ***(14)***. Another recurrent abnormality is der(16)t(1;16)(q21;q11–21), which may also be indicative of a poor prognosis, and which is also found in other malignancies ***(15)***. Anaplastic Wilms' tumors are associated with high hyperdiploidy, typically having >70 chromosomes.

f. Retinoblastoma: This tumor of the eye is most common in children who have inherited a deletion of 13q14.1 and who have subsequently acquired an abnormality of the Rb1 gene on the other chromosome. In most cases, cytogenetic studies have found an isochromosome for the short arms of a chromosome 6, i(6)(p10), sometimes as the sole abnormality.

12. Renal cell tumors: There is an age-related specific cytogenetic abnormality, in that a t(X;1)(p11;q21) is common in tumors occurring in children but is not present in tumors occurring in adults. The implication is that the tumors in children and adults are different diseases, with distinct etiology.

3. Types of Tissue Available for Cytogenetic Studies

It can require considerable cooperation from surgeons, theater staff, and pathologists if a study of cytogenetic abnormalities in solid tumors is to be successful. In many hospitals, tumor material that is surgically removed is immediately placed into formalin or some other fixative, which renders it useless for subsequent cell culture. It is necessary to persuade theatre staff to place at least part of the material into a sterile container without fixative. When the sample arrives in the pathology department, it is usual for the pathologist to have first choice of material for their own studies, with the cytogeneticist being allowed to have whatever is left. By this time, the sample may no longer be sterile. If it is large enough, the best course is to cut out some tissue carefully from the interior, avoiding the contaminated exterior, but avoiding areas where the tumor has become necrotic. If the tissue is not large enough to trim and discard the outer part, then washing two or three times in medium with a high concentration of antibiotics is helpful.

Sometimes the only material available is a fine needle aspirate, which may be inadequate for conventional cytogenetic studies. However, FISH studies can usually be performed on tumor imprints, obtained by gently rolling the biopsy on a clean slide ***(16)***. The adherent cells can be fixed quickly or allowed to air-dry.

Lastly, for tumors that have an accessible surface, such as skin cancers, brush specimens can provide cells suitable for FISH studies ***(17)***.

If fresh tissue is not available, then conventional cytogenetic studies are not possible, and FISH or molecular methods have to be used.

For whatever reason, it is not always possible to obtain suitable primary tumor tissue for genetic and cytogenetic studies. Just occasionally, however, useful results can be obtained from tumor cells that are present in other tissues. These include urine, blood, and serous effusions. The detection of transitional cell tumors using urine samples has already been mentioned. In the author's laboratory there have been two cases of alveolar rhabdomyosarcoma in which cells with a t(2;13) were detected in bone marrow aspirates. (Conversely, however, there have been many other bone marrow aspirates studies from patients with the same disease in which no clonal cells were found. It has to be concluded that a cytogenetic study of a secondary tissue such as bone marrow can occasionally give a positive result, but it is an inefficient assay, and failure to detect a clone is simply uninformative.) A feature of some solid tumors is metastasis to body cavities, resulting in the production of fluids, namely pleural effusions around the lungs or ascites in the abdomen. Such fluids can also be produced as a result of infection and certain nonmalignant conditions, and cytogenetic studies can help to determine whether or not an effusion is malignant ***(18)***.

4. Cytogenetic Studies in the Follow-Up of Patients With Solid Tumors

It is unusual to re-biopsy the same tumor, and so cytogenetic and genetic studies are not used to followup the response to treatment in the same way that they can be for leukemias. However, some tumors do release cells into the circulation, and these can be collected and studied during the course of treatment. Although few such studies have been done so far, initial results from patients with several kinds of cancer have been encouraging ***(19,20)***.

5. Summary

Cytogenetic and genetic studies of solid tumors are an area of continuing rapid growth and discovery. They provide an excellent resource for students interested in research, but more importantly have the potential to benefit patients. As has already happened in the leukemias, identifying certain genetic abnormalities can establish a definite diagnosis and/or can indicate likely response to treatment. Other chromosome abnormalities are ubiquitous, such as rearrangements of chromosome 1, and do not appear to correlate with other clinical features; however, even abnormalities of this sort serve to distinguish between malignancy and reactive conditions. Trisomy 7 is also common in many kinds of tumor and has been found as the sole abnormality in cells from tissues surrounding a tumor; the significance of this is not clear: Either trisomy 7 is associated with some reactive response in normal cells, or else it was a primary clonal abnormality and indicated the presence of infiltrating early malignant cells.

References

1. Tajiri, T., Shono, K., Fujii, Y., et al. (1999) Highly sensitive analysis for N-myc amplification in neuroblastoma based on fluorescence in situ hybridization. *J. Pediatr. Surg.* **34,** 1615–1619.
2. James, L. A. (1999). Comparative genomic hybridization as a tool in tumor cytogenetics. *J. Pathol.* **187,** 385–395.
3. (1995) *Cancer Cytogenetics,* 2nd edit. (Heim and Mitelman, eds.) Alan R. Liss, New York.
4. Slamon, D. J., Clark, G. M., Wong, S. G., Levin, W. J., Ullrich, A., and McGuire, W. L. (1987) Human breast cancer: correlation of relapse and survival with amplification of the HER-2/neu oncogene. *Science* **235,** 177–182.

5. Motzer, R. J., Rodriguez, E., Reuter, V. E., Bosl, G. J., Mazumdar, M., and Chaganti, R. S. K. (1995) Molecular and cytogenetic studies in the diagnosis of patients with poorly differentiated carcinomas of unknown primary site. *J. Clin. Oncol.* **13,** 274–282.
6. Becher, R., Korn, W. M., and Prescher, G. (1997) Use of fluorescence in situ hybridization and comparative genomic hybridization in the cytogenetic analysis of testicular germ cell tumors and uveal melanomas. *Cancer Genet. Cytogenet.* **93,** 2–28.
7. Fiegl, M., Zojer, N., Kaufmann, H., and Drach, J. (1999) Clinical application of molecular cytogenetics in solid tumors. *Onkologie* **22,** 114–120.
8. Alavi, S., Namazie, A., Calcaterra, T. C., Wang, M. B., and Srivatsan, E. S. (1999) Clinical application of fluorescence in situ hybridization for chromosome 11q13 analysis in head and neck cancer. *Laryngoscope* **109,** 874–879.
9. Panagopoulos, I., Mertens, F., Isaksson, M., et al. (2001) Clinical impact of molecular and cytogenetic findings in synovial sarcoma. *Genes Chromosomes Cancer* **31,** 362–372
10. Junker, K., Werner, W., Mueller, C., Ebert, W., Schubert, J., and Claussen, U. (1999) Interphase cytogenetic diagnosis of bladder cancer on cells from urine and bladder washing. *Internatl. J. Oncol.* **22,** 309–313.
11. Zattara-Cannoni, H., Roll, P., Figarella-Branger, D., et al. (2001) Cytogenetic study of six cases of radiation-induced meningiomas. *Cancer Genet. Cytogenet.* **126,** 81–84.
12. McManus, A. P., Gusterson, B. A., Pinkerton, R., and Shipley, J. M. (1996) The molecular pathology of small round-cell tumors—relevance to diagnosis, prognosis and classification. *J. Pathol.* **178,** 116–121.
13. Gordon, T., McManus, A., Anderson, J., Min, T., et al. (2001) Cytogenetic abnormalities in 42 rhabdomyosarcoma: a United Kingdom Cancer Cytogenetics Group Study. *Med. Pediatr. Oncol.* **36,** 259–267.
14. Bown, N., Cotterill, S. J., Roberts, P., et al. (2002) Cytogenetic abnormalities and clinical outcome in Wilms tumor: a study by the U. K. Cancer Cytogenetics Group and the U. K. Children's Cancer Study Group. *Med. Pediatr. Oncol.* **38,** 11–21.
15. McManus, A. P., Min, T., Swansbury, G. J., Gusterson, B. A., Pinkerton, C. R., and Shipley, J. M. (1996) der(16)t(1;16)(q21;q13) as a secondary change in alveolar rhabdomyosarcoma: a case report and review of the literature. *Cancer Genet. Cytogenet.* **87,** 179–181

16. McManus, A. P., Gusterson, B. A., Pinkerton, C. R., and Shipley, J. M. (1995) Diagnosis of Ewing's sarcoma and related tumors by detection of chromosome 22q12 translocations using fluorescence in situ hybridisation on tumor touch imprints. *J. Pathol.* **176,** 137–142.
17. Veltman, J. A., Hopman, A. H. N., Bot, F. J., Ramaekers, F. C. S., and Manni, J. J. (1997) Detection of chromosomal aberrations in cytologic brush specimens from head and neck squamous cell carcinoma. *Cancer* **81,** 309–314.
18. Dewald, G. W., Hicks, G. A., Dines, D. E., and Gordon, H. (1982) Cytogenetic diagnosis of malignant pleural effusions. Culture methods to supplement direct preparations in diagnosis. *Mayo Clin. Proc.* **57,** 488–494.
19. Chen, Z., Morgan, R., Stone, J. F., and Sandberg, A. A. (1994) Appreciation of the significance of cytogenetic and fish analysis of bone marrow in clinical oncology. *Cancer Genet. Cytogenet.* **78,** 1–14.
20. Engel, H., Kleespies, C., Friedrich, J., et al. (1999) Detection of circulating tumor cells in patients with breast or ovarian cancer by molecular cytogenetics. *Br. J. Cancer* **81,** 1165–1173.

11

Human Solid Tumors

Cytogenetic Techniques

Pietro Polito, Paola Dal Cin, Maria Debiec-Rychter, and Anne Hagemeijer

1. Introduction

The field of cytogenetics has had a great impact on many aspects of medical and basic sciences, including clinical genetics and, perhaps most notably, hematology and oncology. Tumor cytogenetics has for many years been dedicated almost exclusively to the study of hematological malignancies, mainly because these are more readily accessible. Important information has been obtained: cytogenetic changes in tumors are acquired, clonal abnormalities that are specifically associated with pathological subtypes of malignancies and premalignant conditions; they have a clear prognostic impact, and are used for choice of treatment, patient stratification, and assessment of minimal disease. Furthermore, molecular investigations of recurrent cytogenetic changes have led to the discovery of genes that play a determining role in leukemia and oncogenesis in general.

Solid tumors are less readily accessible. Pathological and clinical features as well as treatment strategy may interfere and

From: *Methods in Molecular Biology, vol. 220: Cancer Cytogenetics: Methods and Protocols*
Edited by: John Swansbury © Humana Press Inc., Totowa, NJ

limit the possibility of obtaining a sufficient amount of representative tumor cells. Also, obstacles attributable to the biological characteristics of tumor themselves have to be overcome. These difficulties may be stated more specifically: (1) the often low mitotic index in tumors, necessitating their study following relatively long-term culture with the attendant difficulties of successfully culturing the tumor cells; (2) the problem of overgrowth by normal (diploid) stromal cells, particularly in long-term cultures in which the diploid cells have a growth advantage over the sluggish cancer cells; (3) the presence of infection in the tissue sample which can destroy the tumor cells or inhibit their growth in vitro; (4) sampling from a necrotic area of a tumor, thus not supplying an adequate number of viable cells to obtain metaphases for analysis; (5) the acquisition of further chromosome abnormalities during the culturing. Advances in cell culture and in cytogenetic techniques applied to tumor cells have largely obviated some of these difficulties. Starting from the mid-1980s, when the long-term collagenase treatment was utilized in tissue disaggregation ***(1)***, the information about solid tumor cytogenetics has improved rapidly ***(2,3)***. To date, numerous nonrandom chromosomal abnormalities in solid tumors have been described, some of which are specific to certain tumor types. During the years in which cytogenetic techniques were being developed, adapted, and further refined for solid tumors, many differences in methodology existed between laboratories. However, today the protocols are more alike than different, although they should be adapted to the specific needs and environmental conditions of each laboratory. The techniques presented here are in general agreement with the current protocols, and in our hands they have been useful in the study of at least 95% of all solid tumors, allowing cytogenetic analysis of 20–50 metaphases per tumor. This technical procedure is particularly reliable in soft-tissue benign and malignant tumors, in tumors of mesothelial origin, brain tumors, and many epithelial tumors. Certain specific epithelial tumors like breast and prostate are particularly difficult to process successfully. They require specific protocols and culture media, which are extensively described in the original publications ***(4–6)***.

1.1. General Outline of the Technique

A successful cytogenetic analysis of solid tumors is based on a successful culture. The procedure starts in the operating room when the tumor is removed or biopsied. The tissue sample must be representative of the tumor, sterile, and viable (thus not in fixative solutions). Optimally, the tumor biopsy is divided by the pathologist for pathological diagnosis, cytogenetics (culture), and molecular analysis (snap frozen). To shorten the time in culture and avoid overgrowth of fibroblastic stroma, the cultures are incubated in a small culture flask or preferentially directly on microscopic slides mounted in a chamber (Nunc, Naperville, IL).

The cells are incubated at a high density for epithelial tumors and at a low density for mesenchymal tumors. Cell attachment, proliferation, and mitotic rate are monitored by daily examination through an inverted microscope. Time to harvest and duration of colcemid treatment will be determined by this monitoring.

Harvesting and fixation are standard cytogenetic procedures adapted to the condition of the tissue culture, in monolayer, of a very limited amount of cells. Techniques successful for amniotic cell culture are generally applicable.

The different steps to obtain metaphases successfully are discussed in detail, in particular:

- Tumor collection and transport
- Cell disaggregation
- Culture initiation
- Harvesting of culture and metaphases
- Banding techniques
- Freezing of viable cells.

The references cited indicate the original methods from which our current techniques have been adapted and modified.

2. Materials

All solutions and vials have to be sterile, either bought ready to use or prepared in the laboratory and sterilized by autoclaving (salt solutions) or filtration (media and sera).

1. Culture medium: Use either of the media described here, completed by adding the supplements listed:
   a. Dulbecco's modified Eagle medium (DMEM)-F12 high glucose with glutamax (GIBCO). To each 500-mL bottle, add 56 mL of fetal bovine serum (FBS, Hyclone), and 1.2 mL of penicillin–streptomycin (50,000 IU/mL: 50 mg/mL) (Boehringer Mannheim). If necessary, add 1 mL of amphotericin B (Fungizone, 250 µg/mL, GIBCO).
   b. RPMI 1640 or McCoy's 5A: To each 100-mL bottle, add 20 mL FBS (Hyclone), 0.5 mL of penicillin–streptomycin (50,000 IU/mL: 50 mg/mL) (Boehringer Mannheim), and 1 mL 200 m*M* glutamine (Irvine Scientific).

   For transporting samples, use one of these media but without serum and add a double concentration of antibiotics. Dispense into sterile vials and send these to the physicians along with the instruction sheet. The vials can be kept at +4°C for up to 6 mo.
2. Phosphate-buffered saline (PBS) for washing. 500 mL of PBS (without sodium bicarbonate); penicillin–streptomycin (50,000 IU/mL: 50 mg/mL) (Boehringer Mannheim); 2.4 mL and 1 mL amphotericin B (Fungizone, 250 µg/mL GIBCO).
3. Collagenase stock solution: collagenase type 2, 215 U/mg (Worthington Biochemical Inc.) is dissolved in bidistilled water at a final concentration of 2000 U/mL and kept overnight at 4°C. The solution is sterilized by filtration through a 0.2-µm filter. Divide into 1-mL aliquots in small cryotubes. These can be kept for 2–3 mo stored at –20°C. The working concentration of 200 U/mL is prepared immediately before use by adding 1 mL of collagenase to 9 mL of complete medium. Use 10 mL of collagenase solution in a 75-mL tissue culture flask, and 5 mL in a 25-mL tissue culture flask.
4. Colcemid: currently there are commercially available solutions (Karyomax, GIBCO) with the right concentration of colcemid ready to use; alternatively it is possible to use 10 µg/mL of colcemid (GIBCO) to prepare a working solution, diluting it 1:10 in Hanks'

buffered salt solution (BSS), without calcium and magnesium, and storing it in a refrigerator for up to 1 wk.

5. Hypotonic solutions: these may be stored in a refrigerator at 4°C but should be prewarmed to 37°C before use. At the appropriate stage of harvesting, add a volume of hypotonic solution double of the volume of medium present in the flask.
   a. 0.8% Sodium citrate (0.027 *M*) for chamber slides: Dissolve 8 g of trisodium citrate-2-hydrate in 1000 mL of distilled water.
   b. HEPES–EGTA solutions for tissue culture: weigh 4.8 g of *N*-2-hydroxyethylpiperazine-N'-2-ethanesulfonic acid (HEPES) (Sigma Chemical cat. no. NH 9136), 0.2 g of EGTA (ethyleneglycol-*bis*-B-aminoethylethertetraacetic acid) (Sigma Chemical cat. no. NE 4378), and 3.0 g of KCL HEPES-EGTA. Dissolve in distilled water to make 1 L of solution, and adjust the pH to 7.4 using 1 *N* NaOH solution.
6. Methotrexate (MTX) (Lederle) or amethopterin (Sigma Chemicals).
   a. To prepare 10^{-4} *M* stock solution, dissolve 0.954 mg of amethopterin (Sigma) in 20 mL of distilled water or Hanks' BSS. If methotrexate is used (5 mg/2 mL Lederle), dilute 1 mL with 49 mL of distilled water.
   b. Filter to sterilize.
   c. Aliquot into 1-mL portions and freeze for up to 6 mo.
   d. To prepare 10^{-5} *M* working solution, dilute the stock solution 1:10 with distilled water or Hanks' BSS. Sterilize the working solution by filtration.
   e. Aliquot the working solution in 0.5–1-mL portions in small screw-cap tubes and store at –20°C for up to 6 mo.
   f. To use the working solution, thaw at 37°C and add 100 µL per 10 mL of culture medium. Discard any remaining solution. Do not refreeze.
7. Thymidine (Sigma Chemicals): to prepare a working solution (10^{-3} *M*), weigh 2.5 mg of powder into a 15-mL tube and dissolve it in Hanks' BSS. The working solution is sterilized by filtration and aliquoted into 0.5–1.0-mL portions in small screw-cap tubes. The solution can be stored at –20°C for up to 3 mo. To use the working solution, thaw at 37°C and add 100 µL per 10 mL of culture medium.
8. Wright's stain: 1.25 g of Wright's stain (MCB, Manufacturing Chemists, Norwood, OH) is dissolved in 500 mL of methanol, stirred at room temperature for 1 h, and kept at 37°C for 1 d in the dark. The

solution is filtered through Whatman no. 1 filter paper and stored at +4°C in the dark.

9. pH 6.8 Buffer: dissolve one Gurr's pH 6.8 buffer tablet in 1 L of distilled water. May be stored at 4°C. Alternatively, make up phosphate buffer stock solution A (10.68 g of $Na_2HPO_4 \cdot 2H_2O$ in 1 L of distilled water) and phosphate buffer stock solution B (8.10 g of KH_2PO_4 in 1 L of distilled water), and make up the working solution as required by combining 49 mL of A and 51 mL of B, pH 6.8.
10. 0.2 *N* HCl: dilute 16.8 mL of 11.9 *N* HCl in 1 L of distilled water. Keep at room temperature.
11. 20× Saline sodium citrate (SSC): dissolve 175.3 g of NaCI and 88.23 g of trisodium citrate in 900 mL of distilled water. Adjust to pH 7.0 with HCl. Make up to 1 L with distilled water and store at room temperature. Dilute to 2× SSC when it is needed (10 mL of 20× SSC and 90 mL of distilled water).
12. Trypsin for banding (stock solution, 10×): dissolve 1.25 g of Difco 1:250 trypsin in 200 mL of distilled water. Dispense 4 mL into tubes and store frozen for up to 3 mo. Working solution: mix 4.0 mL of stock solution with 36 mL of pH 7.0 buffer in a Coplin jar.
13. 1× Trypsin–EDTA solution (GIBCO).
14. pH 7.0 Buffer: dissolve one Gurr's pH 7.2 buffer tablet and 9 g of NaCI in 1000 mL of distilled water. May be stored at room temperature for up to 3 mo.
15. Giemsa stain: mix 2.0 mL of Harleco Giemsa stain with 48.0 mL of pH 6.8 buffer in a Coplin jar just before use.
16. Freezing medium: culture medium with 20% FBS and 8–10% dimethyl sulfoxide (DMSO) Versene 1:5000 in isotonically buffered saline solution (GIBCO).

3. Methods

3.1. Collection and Transport of Tumor Samples

From 0.5 to 1 g of viable tumor tissue is sufficient in most cases. Smaller biopsy specimens can also be analyzed successfully but they may require prolonged culture.

Sterile, non-necrotic tumor samples are collected in a transport container; this is a sterile tube containing sterile culture medium, an antimycotic, and a double concentration of antibiotics (*see* **Sub-**

heading 2.1.). The transport container can be kept in a refrigerator (4°C) when not in use, and brought to room temperature for collection of the specimens. It should be kept at room temperature thereafter until received in the laboratory.

Use separate containers, correctly and specifically labeled, for multiple samples of one patient, multiple specimens from one tumor, or collection of normal tissue of the same individual. Although tumor samples processed immediately yield better results in tissue culture, successful growth of cells can also be obtained from samples processed up to 24 h after removal. Tumor specimens can be conveniently sent to the laboratory by overnight mail and processed on the following day if they are placed in the correct transport container and left at room temperature.

3.2. Disaggregation of Tumor Tissue (1)

1. Remove the specimen from the transport container and place it in a Petri dish. Before processing, additional washes with PBS (*see* **Subheading 2.2.**) may be necessary, especially for tumors of infected tissue such as lung and gastrointestinal tract. The PBS contains antibiotic and antimycotic agents; we use amphotericin B (250 μg/mL of Fungizone, GIBCO, ready to use). Washes are also indicated when the transport has not been done in a solution containing antibiotic, or when the tumor is necrotic and/or admixed with blood.
2. Do not allow the tissue to dry. Add a few drops of complete culture medium (*see* **Subheading 2.1.**).
3. Mince tissue with sterile curved scissors or surgical blades into fragments 1–2 mm in size and transfer them to a 25-cm^2 tissue culture flask (Falcon).
4. Suspend the fragments in medium containing collagenase at a final concentration of 200 U/mL (*see* **Subheading 2.3.**). If necessary, use culture medium with an antimycotic.
5. Incubate the suspension at 37°C in 5% CO_2 for 16–24 h (overnight).
6. The timing of enzymatic treatment depends on tumor type, but generally overnight incubation is sufficient. Check the progress of disaggregation with the aid of an inverted microscope. A large number of single cells and small clusters of cells may be observed floating at the end of this period.
7. After disaggregation, transfer the cell suspension to a 15-mL conical centrifuge tube.

8. Add PBS + antibiotics (+ antimycotic if necessary) and mix the suspension by pipetting up and down.
9. Centrifuge for 10 min at 1000 rpm (150*g*).
10. Add fresh medium to the flask and reincubate to allow the growth of the cells that were attached to the plastic base during the collagenase treatment.

3.3. Culture Initiation

1. Discard the supernatant and resuspend the pellet in an appropriate amount of medium. Place 0.3–0.5 mL of cell suspension on each of two chamber slides (Nunc). Seed the remainder into flasks.
2. If there is enough pellet, a single-cell suspension can be frozen and stored in liquid nitrogen for future use (*see* **Subheading 3.9.**).
3. Incubate at 37°C in 5% CO_2, to allow the cells to attach.
4. On the following day, remove the medium containing unattached cells and cellular debris from flasks (**Fig. 1A**) and chamber slides and replace it with prewarmed medium. The removed supernatant is used for subsequent interphase fluorescence *in situ* hybridization (FISH) analysis, if required, as it is a source of a significant number of single and original neoplastic cells. Preservation of nuclei requires fixation prior to storage as follows:

   a. Transfer the supernatant to a 15-mL conical centrifuge tube.
   b. Centrifuge for 5 min at 1000 rpm (150*g*).
   c. Discard the supernatant and resuspend the pellet in 10 mL of prewarmed 0.8% (0.027 *M*) sodium citrate (*see* **Subheading 2.5.1.**)
   d. Incubate for 30 min at 37°C.
   e. Centrifuge for 5 min at 1000 rpm (150*g*).
   f. Discard the supernatant and fix the cell suspension with a mixture of methanol and acetic acid (3:1) for 20 min.
   g. Repeat **steps e** and **f** two times.
   h. Store the suspension of nuclei in 1 mL of fixative at –20°C for future use.

3.4. In Vitro Culturing

1. Examine chamber slides and flasks daily through an inverted microscope.
2. Note carefully the mitotic activity of each chamber slide and flask.

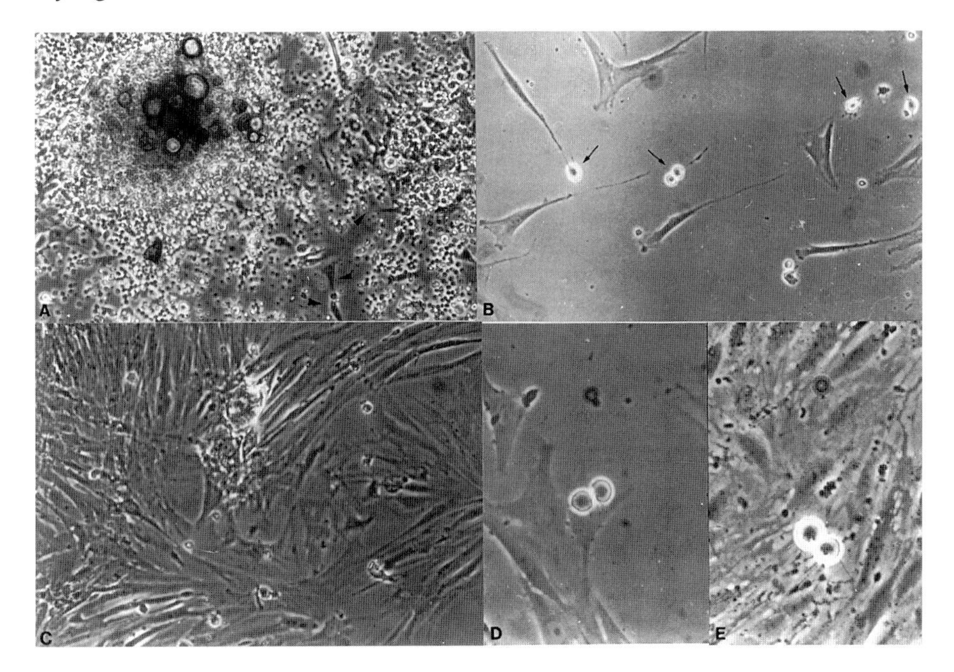

Fig. 1. **(A)** Solid tumor in culture after 16 h. There is cellular debris but also a group of cells attached to the chamber slide (*arrowheads*) (100×). **(B)** Culture after 3 d: *arrows* point to cells in metaphase and in telophase (100×). **(C)** Excessive confluence of the cellular monolayer (100×). **(D, E)** Telophase cell in right confluence **(D)** and in excessive confluence **(E)** (200×).

3. The time for harvest of culture is individualized for each flask or chamber slide and is carried out when "peak" mitotic activity is observed (**Fig. 1B**).
4. In most cases, the chamber slide can be harvested during the first 3–4 d. Significant proliferation of fibroblast-like cells starts by the end of the first week in the flask.

3.5. Harvesting Procedures for chamber slides (In Situ)

1. In a chamber slide the cells are not removed from the growing surface, therefore excessive confluence of the cellular monolayer (**Fig. 1C–E**) must be avoided to allow swelling of the cells during the hypotonic treatment and subsequent spreading of the chromosomes (**Fig. 2A–D**).
2. Colcemid is added to the culture at a final concentration of 0.01 μg/mL (*see* **Subheading 2.4.**) for 16–17 h (overnight).

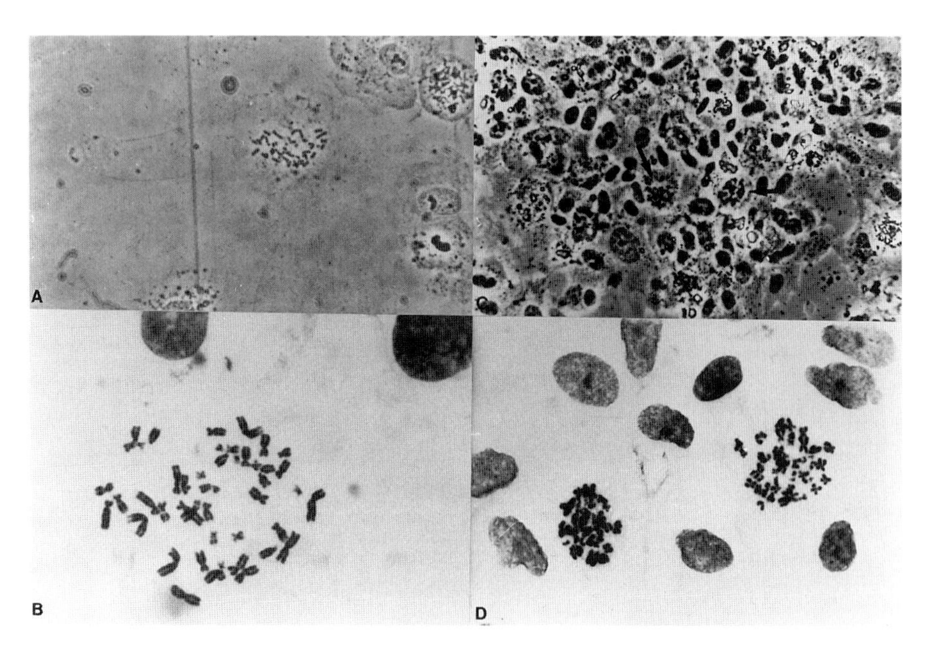

Fig. 2. Metaphase spreads after harvesting before and after G-banding. There is a difference between a metaphase in a "right" confluence (**A** [200×] and **B** [1000×]) and one in an excessive confluence culture (**C** [*arrows* point to metaphase spreads] [200×] and **D** [1000×]).

3.5.1. Manual Harvesting In Situ (7)

1. Carefully remove the medium with a Pasteur pipet.
2. Add approx 2 mL of prewarmed 0.8% (0.027 *M*) sodium citrate (*see* **Subheading 2.5.1.**) slowly down the side of the chamber and allow to stand at 37°C for 20–30 min.
3. Carefully add 1 mL of *cold* freshly prepared fixative methanol–acetic acid (3:1) to the hypotonic solution.
4. After 2 min, remove all fluid.
5. Add 2 mL of fresh cold fixative slowly down the side of the chamber.
6. After 20 min, remove the fixative.
7. Repeat **steps 5** and **6** twice with the final fixation lasting for only 10 min.
8. Remove fixative and air-dry the chamber slides at room temperature. Check spreading under phase-contrast microscopy.

3.5.2. Automatic Harvesting

Recently some modern automatic processors have become available to perform *in situ* harvesting. The processor can remove and add solutions with a robotic arm moving in three dimensions, controlled by a computer. Many of these automatic sample processors are commercially available; in our laboratory we use the system Tecan RSP 5000 (Tecan AG Switzerland) for *in situ* harvesting of amniotic cell cultures and solid tumor biopsies. At the end of the automated procedures the usual methods of banding can be utilized.

3.5.3. Harvesting Procedures for Flasks (8)

Following a careful monitoring of the mitotic activity by an inverted microscope, three harvesting methods can be employed that differ in colcemid concentration and incubation time.

3.5.3.1. Short-Term Colcemid Treatment.

1. Colcemid (*see* **Subheading 2.4.**) is added to the culture at a final concentration of 0.02 µg/mL.
2. Harvest 3–4 h later.

3.5.3.2. Prolonged Colcemid Treatment.

1. Colcemid is added to the culture at a final concentration of 0.01 µg/mL.
2. Harvest the next morning.

3.5.3.3. MTX Synchronization Procedure *(9)*.

1. MTX (*see* **Subheading 2.6.**) is added at a final concentration of 10^{-7} *M* (10 µL/mL) overnight (17 h).
2. The following morning, remove and discard the medium containing MTX.
3. Wash once with fresh prewarmed medium, then discard this medium.
4. Add fresh complete medium supplemented with thymidine (*see* **Subheading 2.7.**) at a final concentration of 10^{-5} *M* (10 µL/mL).
5. Incubate for 4–5 h.
6. Colcemid is added at a final concentration of 0.01 µg/mL.
7. Harvest 1–2 h later.

3.6. With All Three Methods, the Subsequent Steps of the Harvesting Procedure Are the Same

1. Discard the medium.
2. Add directly to the flask a volume of prewarmed hypotonic solution (*see* **Subheading 2.5.2.**) at least twice that of the previously discarded medium, for example, if the discarded medium was 10 mL, replace by 20 mL of hypotonic solution.
3. Incubate for 30 min at 37°C.
4. Using a cell scraper, mechanically remove the remaining attached cells.
5. Transfer the cell suspension to a 50-mL conical centrifuge tube.
6. Centrifuge for 5 min at 1000 rpm (150*g*).
7. Discard the supernatant and fix the cell suspension with a mixture of methanol and acetic acid (3:1) for 20 min.
7. Repeat **steps 6** and **7** four times.
8. Store the cell suspension in an appropriate volume of fixative (10–15 mL) overnight at –4°C.

3.7. Preparation of Slides from Flask Harvest

To obtain good spreading of the chromosomes, the relative humidity of the environment should be approx 60%.

1. The slides should be carefully cleaned and degreased; leave the slides overnight in 20% glacial acetic acid, rinse in denatured ethanol, and wipe them with a dust-free cloth.
2. Drop the cell suspension onto these clean slides, moistened with breath, gently rocking them and blowing on them.
3. Place the slides upright on filter paper to remove the excess of fluid.
4. Place the slides on a hotplate (~ 50°C) and allow them to dry (~ 1 min).
5. Check the quality of preparation with an inverted microscope. If difficulties are experienced in spreading the metaphases, other methods can be tried such as dropping the suspension from 2–4 ft on cold, wet slides (moistened with water), and flaming the slides.
6. Store the pellet eventually left in an Eppendorf tube in 1 mL of fresh fixative at –20°C for future use (*in situ* hybridization, microdissection, etc.).

3.8. Banding and Staining Procedure

There are many techniques available to obtain a banding pattern on the chromosomes, but the most widespread and simply performed for routine use are the following three resulting in a G-banding pattern. All three require the aging of slides for 20 min in a 100°C oven.

3.8.1. G-Banding with Wright's Stain (10)

1. Stain the slides directly with 1 mL of 0.25% Wright's stain (*see* **Subheading 2.8.**) mixed with 3 mL of phosphate buffer, pH 6.8 (*see* **Subheading 2.9.**), for 2 min. Prepare this mixture freshly.
2. Rinse quickly under tap water.
3. Check the quality of the staining procedures under a microscope, preferably without using immersion oil. If the banding pattern appears too dark or too pale it is possible to destain and/or restain the slide. If immersion oil has been used, it is necessary to clean the slides by dipping them in xylene for 5–10 min before following the destaining procedures. Once the slides are dried from the xylene, immerse them in absolute methanol for 2 or 3 min. Stain again for a shorter or longer time accordingly.

3.8.2. C-G Banding (11)

This protocol often offers the best results after the automatic harvesting procedures:

1. Incubate the slide in 0.2 *N* HCl (*see* **Subheading 2.10.**) for 5 min.
2. Rinse very well under tap water, and dry with a hair dryer.
3. Incubate in 2× SSC solution (*see* **Subheading 2.11.**) at 60°C for 20 min.
4. Rinse very well under tap water and dry with a hair dryer.
5. Stain the slides with 1 mL of 0.25% Wright's stain (*see* **Subheading 2.8.**) mixed with 3 mL of phosphate buffer, pH 6.8 (*see* **Subheading 2.9.**), for 2 min. Time variations may result in darker or paler slides, as in the previously described methods, but generally with this pretreatment the Wright's stain is more stable.
6. Rinse quickly under tap water.
7. This technique gives G- and C- bands. If the C-banding is too prominent it is necessary to modify the time of SSC exposure to obtain G-bands as well.

3.8.3. G-Banding with Trypsin and Giemsa (12)

1. Treat slides with 0.025% trypsin in pH 7.0 buffer for between 5 s and 3 min (*see* **Subheadings 2.12.** and **2.14.**).
2. Rinse in pH 7.0 buffer for 5 s.
3. Rinse in second pH 7.0 buffer for 1 min.
4. Stain in Giemsa (*see* **Subheading 2.15.**) for 2 min.
5. Rinse with distilled water.
6. Dry gently in warm air (hair dryer).

3.9. Freezing Procedures

The freezing and storage of cancer cells before and/or after culture is of utmost importance for subsequent molecular studies as well as for *in situ* hybridization, microdissection, and so forth.

3.9.1. Single-Cell Suspension

If there are enough cells after the collagenase treatment, it is advisable to freeze some of them.

1. Resuspend the cell pellet in 2 mL of freezing medium containing serum and DMSO (*see* **Subheading 2.16.**).
2. Fill the cryogenic ampoules (Sterilin) and close them by melting.
3. Start the freezing program (*see* the last paragraph in **Subheading 3.9.2.**).

3.9.2. Cultured Cells

Cells that are growing in a flask and are not to be used for cytogenetic harvesting need to be frozen after about 1 wk in culture, before stromal cells overgrow.

1. Remove the medium from the flasks.
2. Wash twice with approx 2 mL of Versene 1:5000 in an isotonically buffered saline solution for 30 s each time.
3. Add 2 mL of 1× trypsin–EDTA solution.
4. Incubate at 37°C for 10–15 min.
5. Detach the cells from the bottom of the flask.
6. Incubate again at 37°C.

7. Add culture medium to stop the action of the trypsin, and transfer to a conical tube.
8. Centrifuge at 1000 rpm (150*g*) for 10 min.
9. Discard the supernatant and resuspend the pellet in 2 mL of freezing medium (*see* **Subheading 2.16.**).
10. Fill the cryogenic ampoules (Sterilin) and close them by melting.
11. Start the freezing program.

At the end of both procedures the cells need to be placed in an automated freezing device, commercially available, that allows them to reach the storage temperature in a programmed and computer-driven manner. These devices provide an exact cooling rate of 1°C/min down to 0°C. Then the machine supplies a rapid flush of gas phase nitrogen to obtain a subsequent cooling rate of 4°C/min until the storage temperature is reached. The cells can then be cryopreserved in liquid nitrogen. This method, minimizing the damage caused by intra- and extracellular ice formation, improves cell rescue at thawing. Alternatively, the vials with the cells can be placed in the gas phase of liquid nitrogen (usually in a specific holder placed in the top of the liquid nitrogen container) at a cooling rate of 1 or 2°C/min. After 1 h the vials can be immersed in liquid nitrogen for storage.

References

1. Limon, J., Dal Cin, P., and Sandberg, A. A. (1986) Application of long-term collagenase disaggregation for the cytogenetic analysis of human solid tumors. *Cancer Genet. Cytogenet.* **23,** 305–313.
2. Sandberg, A. A. (1990) The *Chromosomes in Human Cancer and Leukemia,* 2nd edit., Elsevier, Amsterdam.
3. Mitelman, F., Johansson, B., and Mertens, F. (eds.) (2002) Mitelman Database of Chromosome Aberrations in Cancer. http://cgap.nci.nih.gov/Chromosomes/Mitelman <http://cgap.nci.nih.gov/Chromosomes/Mitelman>.
4. Pandis, N., Heim, S., Bardi, G., Limon, J., Mandahl, N., and Mitelman, F. (1992) Improved technique for short-term culture and cytogenetic analysis of human breast cancer. *Genes Chromosomes Cancer* **5,** 14–20.

5. Peehl, D. M. and Stamey, T. A. (1985) Growth response of normal, benign hyperplastic and malignant human prostatic epithelial cells in vitro to cholera toxin, pituitary extracts, and hydrocortisone. *Prostate* **8,** 51–61.
6. Lundgren, R., Mandahl, N., Heim, S., Limon, J., Hendrikson, H., and Mitelman, F. (1992) Cytogenetic analysis of 57 primary prostatic adenocarcinomas. *Genes Chromosomes Cancer* **4,** 16–24.
7. Peakman, D. C., Moreton, F. M. F., and Robinson, A. (1977) Prenatal diagnosis: techniques used to help in ruling out maternal contamination. *J. Med. Genet.* **14,** 37–39.
8. Gibas, L. M., Gibas, Z., and Sandberg, A. A. (1984) Technical aspects of cytogenetic analysis of human solid tumors. *Karyogram* **10,** 25–27.
9. Yunis, J. J. (1976) High resolution of human chromosomes. *Science* **191,** 1268–1270.
10. Yunis, J. J. (1981) New chromosome techniques in the study of human neoplasia. *Hum. Pathol.* **12,** 540–549.
11. De La Mala, M. and Sanchez, O. (1976) Simultaneous G- and C-banding of human chromosomes. *J. Med. Genet.* **13,** 235–236.
12. Seabright, M. (1971) A rapid banding technique for human chromosomes. *Lancet* **2,** 971–972.

12

Analysis and Interpretation of Cytogenetic Findings in Malignancy

John Swansbury

The preceding chapters have described the processes involved in getting metaphase divisions from a sample onto a slide, ready for analysis. Time, skill, and experience generously spent on these techniques will do much to make the next stages easier: the analysis and interpretation of these metaphases. Much of the following subheadings provide practical advice about these important aspects of a cytogeneticist's work. They are mainly derived from experience with hematologic malignancies, as these have formed the bulk of most malignancy cytogenetics work so far, and have become the subject of professional guidelines drawn up to ensure that consistent and uniformly high standards of analysis are applied in all laboratories. Much the same principles are now being applied to the analysis of solid tumors.

1. Microscopy: Good Practice

It is likely that the cytogeneticist will spend many hours seated at the microscope, and it is important that factors such as lighting and the chair height are adjusted to minimize any risks of injury to eyes or the back caused by strain or long-term poor posture. In some

From: *Methods in Molecular Biology, vol. 220: Cancer Cytogenetics: Methods and Protocols*
Edited by: John Swansbury © Humana Press Inc., Totowa, NJ

centers the amount of time spent at the microscope is limited to a maximum of 4 h a day, and it is mandatory to have a 5–10 min break every hour.

A 10× objective lens can be used for screening, and an oil-immersion 100× objective is usually required for studying the metaphases. Most people find it easiest to screen across the slide as if they were reading, like this:

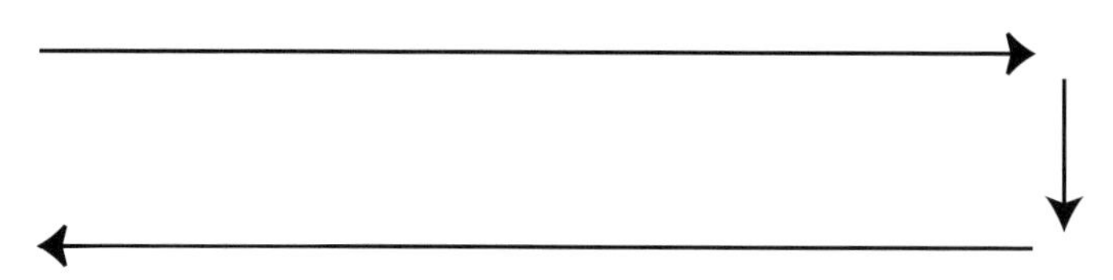

When there are few divisions available, however, a better way of systematically screening the slide is up-and-down, moving across the slide in small increments:

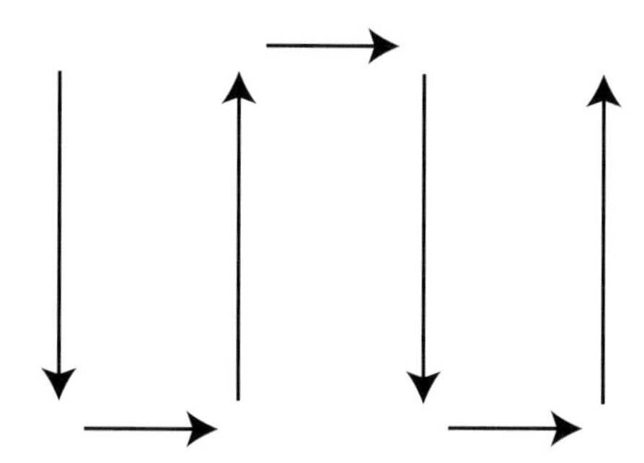

When selecting divisions, remember to bias toward those with poor morphology. Selection of only good mitoses can lead to failure to detect the presence of an abnormal clone. If the quality of metaphases is poor then full analysis may not always be possible. However, even if the chromosomes can only be counted and/or grouped, useful information can sometimes be obtained.

The analysis of polyploid mitoses may appear daunting but in some cases their morphology is better and so abnormalities may be more apparent. Conversely, do not disregard cells that obviously have few chromosomes, assuming that they are just broken with random loss: Clones with a near-haploid complement, for example,

26–29 chromosomes, are known to occur in acute lymphoblastic leukemia (ALL). These can be distinguished from cells with random loss because there will be at least one copy of every chromosome.

2. Chromosome Analysis

Direct analysis down the microscope is possible with experience, and is generally adequate when full analysis is not required (e.g., because the sample is just being screened for a previously identified abnormality) or when any clonal abnormality is simple. It can help to make a sketch of the positions of the chromosomes on some rough paper. Direct analysis is probably the most rapid way of working through a case. However, there are drawbacks. For example, it requires considerable experience; it requires a high level of alertness; it is easy to detect an obvious abnormality and miss a coexisting subtle one; checking involves a complete reanalysis; and no long-term record of the findings is kept (most centers find that slides tend to deteriorate and become unusable).

Preparing a karyogram (the formal arrangement of chromosomes) of each division analyzed produces the most reliable and easily checked analysis. To do this it used to be necessary to take photographs of each metaphase, process and print the film, then cut out the chromosomes and stick them onto a karyotype card, similar to that shown in **Fig. 1**. This took many hours to do, more than most laboratories could spare. However, there are now several computer-based, semi-automated systems that can make a digitized image of the metaphase and help to produce the karyogram within a few minutes. Although such computer-based systems are expensive, their use greatly increases the confidence a laboratory has in the accuracy of the analysis of its cytogeneticists.

Most centers expect to analyze 20–25 divisions for a diagnostic study, unless a clone can be adequately defined with fewer.

When working on follow-up samples for a previously found abnormality, it is often sufficient to check for the presence/absence of the abnormality. However, a few divisions should still be fully analyzed, particularly if the patient is in apparent relapse more than

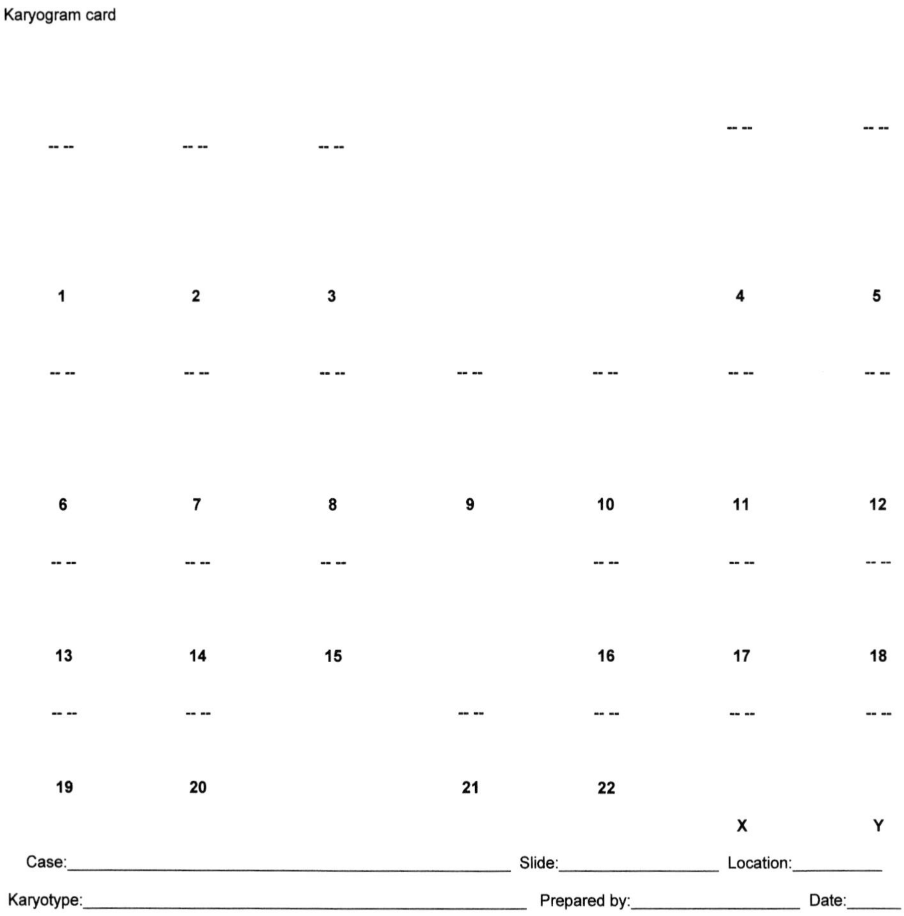

Karyogram card

1 2 3 4 5

6 7 8 9 10 11 12

13 14 15 16 17 18

19 20 21 22 X Y

Case:________ Slide:________ Location:________

Karyotype:________ Prepared by:________ Date:________

Fig. 1. Karyogram card. A photograph of a metaphase spread is cut up and the chromosomes are stuck onto the card like this, with the centromeres aligned on the dashes.

a year after diagnosis. A new clone may be present, which may indicate that the patient has not relapsed but has a new, secondary malignancy.

If the case being studied has a complex and variable clone, it can be helpful to summarize the findings on an analysis sheet such as that shown in **Fig. 2**.

3. Reporting Times

A clinician will often want a quick result to help with the management of the patient, and may not appreciate that it takes time to accumulate divisions, especially from stimulated or long-term cultures, in addition to the time needed for harvesting and processing. The result itself can also have an effect on reporting time, as it takes less time to analyze and report five divisions if they are all from a simple clone than to analyze 25 divisions with no abnormality. In some circumstances a useful result can be provided within 24 h if it is known at the time the sample is received that an urgent result is wanted. It may also be necessary to consider having concurrent fluorescence *in situ* hybridization (FISH) and cytogenetic studies in case one or other does not provide a result. At present, this applies particularly to confirming the diagnosis of acute promyelocytic leukemia, so that treatment with all-*trans*-retinoic acid, which is specific for this disease, can be started promptly.

It is not wise to promise to give an urgent result within a time that is too short to allow for failure of a particular culture, or for checking of the findings by a second cytogeneticist.

Few laboratories can provide such a quick turnaround time in routine practice, but all laboratories should endeavor to report studies promptly enough to be useful when making clinical decisions. The time may vary according to local practice. For example, if in a particular center all patients with acute myeloid leukemia (AML) are given a standard course of induction chemotherapy that is not altered until a d 28 assessment is made of the response, then for this clinical decision all cytogenetic studies of diagnostic AMLs should be reported in less than the 28 d. However, there is more to the clinical management of a patient than choosing treatment; the patient and his or her family will want to know urgently what is the prognosis, and for this reason as much as any other, studies should be reported as soon as possible.

There may be guide maximum reporting times given by national quality assessment schemes. In the United Kingdom, for example, the

Karyotype analysis summary of complex/variable clone

Case Name: ______________________ Date: __________ Analyst: ______________

Slide / Verniers or cell No.	1	2	3	4	5	6	7	8	9	10	11	12	13	14	15	16	17	18	19	20	21	22	XY	mar	N°

Summary:

guide reporting time is 14 d for all diagnostic studies of leukemia and 21 d for myelodysplastic syndromes and other hematologic disorders.

4. Quality Control

The success or failure of cultures is also dependent on the quality of the reagents used. Therefore, any unexplained change in success rate should be investigated promptly. For this purpose, it is essential to keep records of the manufacturer's batch numbers of serum, culture media, and reagents, and also the date when they were brought into use in the laboratory. Record should also be kept of the person responsible for the processing of each sample. All these should help to discover the cause of repeated failures, as well as helping to identify ways of improving the quality of the preparations.

Every laboratory needs to have a clearly defined system in place to ensure that the results it provides are accurate. In many countries it is now required that there is documented evidence of such systems, and a record kept of any errors that are discovered.

4.1. Ensuring that a Sample Is Assigned to the Right Patient

Assess every stage of every procedure and search for ways that errors could occur (*see* **Table 1**). This is to minimize any possibility of the sample, analysis, or report being assigned to the wrong patient. Some laboratories have a rule that staff must not talk to each other while setting up cultures or labeling slides, to minimize distractions that could result in mistakes.

In the author's laboratory, once a new sample has been recorded on the laboratory computer database, adhesive labels for culture tubes and slides can be printed out automatically with the patient's

Fig. 2. (opposite page) A table like this is useful when the case being analysed is complex, variable, or not fully analyzable. One row is used for each metaphase. In each box is entered whether the particual chromosome is gained, lost, or abnormal. The number of markers (unidentifiable chromosomes) and the total number of chromosomes are entered in the last two boxes

Table 1
Quality Control:
Ensuring that a Sample Is Assigned to the Right Patient

Stage	Action
Receipt of sample	Check that the details on the container match those on the referral form.
Recording sample in day book or on computer	Check whether or not the patient has been studied previously. Check that no other patients with the same name have been studied. If the case is given a laboratory number, check that it is entered correctly.
Setting up cultures	Set up all cultures for one case at a time. Label all the culture tubes, and then check again that the labeling matches the patient's ID on the sample container.
Spreading slides	Spread one culture (or one case) at a time. Label all the slides before spreading, if possible, or else immediately afterwards. Do not leave the labeling until later! Check again that the slide label matches the patient's ID on the culture tube.

ID, the date, the type of culture, and any other necessary information. Other laboratories use labels with bar codes.

4.2. Ensuring that the Analysis Is Correct

Analysis directly down the microscope is possible with experience but at least one metaphase cell from each case should be photographed, and preferably made into a karyogram, as a record. Ideally, all the metaphases analyzed are fully karyotyped but this is usually possible only if a computerized karyotyping system is available.

Another cytogeneticist must analyze at least one division. This is particularly important in cases where no clone has been found. Such a procedure is obviously much easier if formal karyograms have been prepared, photographically or electronically, but should never be omitted even if direct analysis down the microscope is used, despite the extra work involved.

Finally, check that the karyotype is consistent with the patient's details as described on the referral letter. For example, does the XX or XY status match what is expected from the gender of the patient—or of the donor if the study is post-transplant?

4.3. Ensuring that the Final Result Is Consistent with the Reason for Referral

Samples are often sent with only a provisional diagnosis, and sometimes it is the result of the cytogenetic study that indicates more precisely what the diagnosis is. However, for each diagnostic study there should be a formal follow-up system to obtain a confirmed final diagnosis from the clinician. This is particularly important when there has been a major change, for example, when immunophenotyping results indicate that a lymphoid malignancy was of T-cell origin, which may require the analysis of phytohemagglutin (PHA)-stimulated divisions.

One way of doing this follow-up is to send with each report a form to the clinician, asking for it to be returned with details of the final diagnosis.

It is also important to keep sufficient records of analysis to be able to provide evidence to support a report if it is queried later, and to be able to review the material analyzed if subsequent studies provide further information. Slides should be kept in the dark to avoid fading. How long they are kept for depends on the policy of the laboratory or on the local legal requirements. Some laboratories do not have enough storage space to keep slides indefinitely and discard them after 2 yr. In the author's laboratory, in which most of the samples are from patients with leukemia, the slides are kept for at least 7 yr, in case there is a relapse of the leukemia or the emergence of a secondary leukemia. In addition, all spare fixed cells are stored at –20°C in case they are needed for subsequent FISH studies or research.

4.4. Participating in External Quality Control Audits

In many countries there are voluntary or compulsory schemes in which annual figures and/or examples of a laboratory's work are submitted for external assessment. Although taking part is time consuming, these schemes help to ensure that laboratory standards are maintained. A satisfactory assessment is an independent indication that the laboratory is providing a good service. This is measured in terms of success/failure rates, clone detection rates for selected diagnoses, the time between receiving a sample and issuing a report, achieving minimum standards of chromosome quality appropriate for the type of analysis, providing a report that accurately uses the International System for Human Cytogenetic Nomenclature *(1)*, and describing the findings in a away that is useful to the clinician.

It is recommended that every laboratory participates in a scheme of this type. If there is no national scheme available where the laboratory is located, it is possible to join a scheme in another country.

5. The Interpretation of Cytogenetic Findings

Cytogenetics is a relatively new discipline in pathology, having become established for routine use in little over 20 yr. New clinical associations are still being discovered and new applications are being developed. It is rare for the clinician who treats a patient to be actively involved in the work of a cytogenetics laboratory; most clinicians and oncologists will have only a limited grasp of the appropriate use of cytogenetic studies of malignancy, and may not be fully aware of the significance of the results. Therefore it is usually not enough that a cytogeneticist merely reports the findings of a cytogenetic study; it is often necessary to know the relevance and implications of those findings, so that the clinician can be advised on how to use them to the patient's greatest advantage.

For accurate communication with clinicians, a cytogeneticist should have an understanding of:

1. The appropriate use of cytogenetic studies.
2. The classifications of malignant and related premalignant disorders.

3. Their common etiology and clinical course.
4. The types of treatment used.
5. The possible effects of cytogenetic results on choice of treatment.
6. The ways that cytogenetics can assess response to treatment.

The cytogeneticist should develop a good working relationship with the clinicians who use the laboratory's services, so as to be aware of the level of explanation that may be required. In some centers, for example, it is the policy that the cytogenetics report does not contain any opinion about the prognostic associations of any clonal abnormality that has been found. In other centers, this is accepted as being helpful. For example, a clinician may not be aware how the mortality associated with a particular chromosome abnormality compares with the morbidity associated with intensive treatment, such as bone marrow transplantation (BMT). In the absence of other, disadvantageous, factors, it might be regarded as unethical to subject a patient with a good-risk abnormality such as t(8;21) to the high risk of undergoing BMT. The balance of these risks may change in time: as advances in treatment are made, so the cytogeneticist also needs to keep up-to-date with how patients are managed. This is not to overstate the role of the cytogeneticist in providing advice: The clinician will know the results of all the other laboratory investigations into the patient's disease, and has ultimate responsibility for the treatment and management of the patient.

As the cytogeneticist's experience grows, so will the confidence in reporting and interpreting the results of cytogenetics studies. For the inexperienced, however, the following sections may be helpful in avoiding some of the more common misunderstandings.

5.1. Inherited/Acquired Abnormalities

Some clinicians may fail to appreciate the difference between acquired and inherited cytogenetic abnormalities, and the type may need to be stated clearly when reporting the results of a cytogenetic study.

5.1.1. Inherited (Constitutional) Abnormalities

These are usually present in all the cells of an individual. In most cases, gain or loss of a whole chromosome, or part of a chromosome, has a major effect on that person's physique, which the clinician should mention when referring a sample for cytogenetic study. Some people have a mosaic constitutional karyotype, such that the abnormality is not present in all the cells of an individual. Depending on when the abnormality arose during embryogenesis, different body tissues may or may not have the abnormality, and the effects of the abnormality can be minimal. Also, some people do have constitutional balanced translocations that have no apparent physical affect and so may be unsuspected. This possibility should be considered whenever a cytogenetic study discovers an abnormality that is present in all divisions and that does not fit the stated diagnosis (*see* **Note 5** in Chapter 9). In such cases, it is important that further studies are made, for example, using PHA-stimulated lymphocytes or a skin biopsy, to check that this abnormality is not in fact constitutional.

There are also some inherited chromosome fragility syndromes, such as Fanconi's anemia, Bloom's syndrome, and ataxia telangiectasia, in which there is an inherited inability to repair DNA damage. This inability may in turn predispose to developing some kinds of malignancy. In Fanconi's anemia and Bloom's syndrome, the chromosome abnormalities are usually clearly random damage; in ataxia telangiectasia, however, certain chromosome loci (e.g., 14q11) are preferentially involved and these can form recurrent translocations that give the impression of deriving from a clone.

The chromosomes of most people have a tendency to break at one or more specific locations, known as fragile sites, which are inherited. These are not usually detected unless the cells are cultured in particular ways. For example, many fragile sites need a folate-deficient culture medium before they start to appear. There has been very little evidence to suggest that any fragile sites have a direct connection with malignancy.

5.1.2. Acquired Abnormalities

These arise during the lifetime of an individual and occur in one cell; if that cell is still capable of dividing, then some abnormalities can be passed on to the progeny of that cell. Most acquired abnormalities are random, reactive, or clonal.

Damage caused by exposure to agents such as radiation and certain chemicals is usually clearly nonclonal, with the breaks occurring apparently randomly distributed around the karyotype. These are frequently seen in, for example, patients who have started chemotherapy or radiotherapy for their malignancy. Every cell usually has a different abnormality, although certain chromosomes may be more frequently involved, such as 7 and 14.

Because most people are exposed to low levels of harmful substances in the environment, it is quite common to detect a low frequency of random abnormalities in the stimulated lymphocytes of normal people. It is less common to find random abnormalities in bone marrow cells, as most cells with damage die at the next cell division, and bone marrow cells have a rapid turnover. In the author's laboratory, low but above-normal levels of random damage have been seen in the bone marrow cells of some children with ALL at diagnosis, with no obvious cause. If the frequency is above 3–5%, and there is no obvious recent exposure history, then the possibility of the patient having a chromosome fragility syndrome should be considered and further investigations should be made, as patients with these syndromes may have greatly increased sensitivity to treatment, particularly radiotherapy.

Other commonplace exposures, such as smoking, can also result in chromosome damage. A viral infection may sometimes be indicated by the presence of divisions with large numbers of tiny fragments. However, this should not be confused with the finding in some kinds of malignancy, when gene amplification can result in the presence of paired "double minutes." These can vary in number from one or two per cell up to over a hundred.

Not all acquired abnormalities are associated with malignancy, even when they appear to be, or are, clonal. Reactive lymphocytes in normal kidney tissue around renal carcinoma have been found to have trisomy for chromosome 5, 7, 10, or 18, or loss of the Y *(2)*. Also, infants with Down syndrome are prone to have a "leukemoid reaction" or transient abnormal myelopoiesis (TAM) at birth, which can closely resemble leukemia *(3)*, but that resolves spontaneously, even when a clone is present. The most common clonal abnormalities are extra copies of chromosome 21, and abnormalities of chromosomes 5 or 7. This usually benign condition, TAM, must not be confused with congenital leukemia, in which t(4;11)(q21;q23) is the most common abnormality, and which has a very poor prognosis.

More usually, however, a consistent abnormality occurring in a population of cells means that they are from a clone and in most disorders this equates with the presence of a malignancy.

There has been no ISCN definition of a complex clone. The author regards a simple clone as having one gain, loss, or translocation, and a complex clone as having more than one abnormalityThe generally accepted definition of a clone is two cells with the same gained chromosome or the same structural abnormality, or three cells with the same chromosome lost. These definitions must be used in the correct context: Some patients have a more fragile cell membrane that ruptures easily during slide spreading, resulting in random loss of chromosomes. If this is prominent in a particular study, then it would be necessary to find significantly more than three cells with the same lost chromosome before being confident that there really was a clone present.

5.1.3. Abnormalities that Can Be Acquired and Inherited

There are some abnormalities that can occur in both circumstances. For example, trisomy 21 is common both as an inherited abnormality (in Down syndrome) and as an acquired abnormality (in several kinds of hematologic malignancy); it is necessary to consider both possibilities if a +21 is found in a cytogenetic study. Most individuals with Down syndrome have characteristic physical features.

Other abnormalities are less likely to cause confusion; trisomy 8, for example, is one of the most common acquired abnormalities in AML, but it also occurs, rarely, as an inherited disorder *(4)*. However, most fetuses with trisomy 8 die before birth; some that are mosaic do survive but generally there are severe physical deformities, and these should serve to identify such individuals.

Loss of the Y chromosome occurs at increasing frequency in men over the age of 50, and is almost always an age-related effect, probably not an indication of the presence of a clone ***(5)***. However, rare cases with a clone having –Y as the sole abnormality have been reported. If the significance of a population of cells at diagnosis with –Y is uncertain, it may be necessary to ask for a remission sample: If the –Y cells are still present, then they are probably not clonal. Although less common, age-related loss of an X chromosome also occurs in women. Note that loss of an X (in females) or the Y (in males) as an abnormality secondary to t(8;21)(q22;q22) is very common in AML, and is then part of the clone.

5.1.4. Distinguishing Between Abnormality and Normal Variation

The heterochromatic regions just below the centromeres of chromosomes 1, 9, and 16 can have wide variation in size between individuals and can sometimes look abnormal. Illustrations are given in Chapter 18. Checking a few divisions from a PHA-stimulated culture should confirm that most of any unusual appearances of these regions are simply part of the patient's inherited constitutional karyotype. It can be difficult to be sure of a suspected del(16)(q22) in a patient whose 16s have widely different heterochromatin.

A pericentric inversion of chromosome 9 occurs in the constitutional karyotype of up to 10% of the population without any known clinical effect. An illustration is given in Chapter 18.

These variations are inherited, and so are sometimes useful after a bone marrow transplant in determining the origin of divisions when the host and donor are of the same sex.

See also Chapter 18 **Subheading 6.**

6. Sources of Information about Chromosome Abnormalities

Cytogenetics of malignancy is a discipline in which there is still rapid progress and reporting of new clinical associations. Some of the more common malignancy-associated cytogenetic abnormalities are mentioned in this book; for more details, *Cancer Cytogenetics* ***(6)*** is highly recommended. However, for novel or rare abnormalities, it will be necessary to undertake a literature search. *Cancer Genetics and Cytogenetics* has long been a specialist journal for the publication of the chromosome abnormalities found in malignancy. Other journals that often have papers relevant to malignancy cytogenetics include *Genes, Chromosomes and Cancer*, *Blood*, the *British Journal of Haematology*, and *Leukemia and Lymphoma*.

However, the most convenient way of undertaking searches is usually via the Internet. Many medical and scientific journals are published in an electronic format. For older publications, there are now electronic collections of medical and scientific journals, usually with just the abstract rather than the whole text. In addition, there is an on-line collection of all published malignancy cytogenetic abnormalities at *cgap.nci.nih.gov/Chromosomes/Mitelman*. This is the latest form of a series of editions of the *Catalog of Chromosome Aberrations in Cancer* produced by Professor Felix Mitelman of the University of Lund, Sweden, which have long been an essential reference for malignancy cytogeneticists (*see* **Fig. 1** in Chapter 1).

Do not accept everything that is published as infallible. Mistakes are sometimes made, some conclusions are unjustified, and some conclusions are subsequently overturned by advances in technology or treatment. Always be alert to the following:

1. Be cautious about using conclusions that are drawn from small series of cases, or those that are not confirmed by similar findings made in at least one other center. A small group of patients from one center may have had a biased ascertainment, or may simply not be representative of patients elsewhere, so the conclusions drawn from them may not apply to patients in other centers. If there has been no subsequent confirmation, it may be because other centers cannot reproduce the results. It is often more difficult to publish negative results, even though these may be more important.
2. Be cautious about using data in older publications: Sometimes the diagnoses were based on outdated classification systems, the cytogenetic data will not have been supported by FISH and molecular studies, and survival data may not be so good because the treatments

used were not as good as those in more recent protocols. Also, remember that prognostic effects associated with particular chromosome abnormalities may change as improvements in treatment are introduced.

7. Treatment Status in Acute Leukemia

Some clinicians do not understand that if the patient has acute leukemia and has had any cytotoxic or steroid therapy, then it is highly probable that any clone will rapidly become undetectable, and so a normal result will be obtained. This is true even if a morphological assessment of the bone marrow finds persistent disease. Therefore, to be of value for a diagnostic study, the sample for cytogenetic study must be taken before starting any treatment.

Sometimes clinicians wish to send samples taken at d 7 or d 14 of treatment. In our experience, the bone marrow activity is usually severely suppressed and either no divisions are obtained, or else they are all normal. Rare exceptions occur: In one case studied in the author's laboratory, a pretreatment sample produced only normal analysable divisions, but a trisomy 13 clone was detected in a sample taken shortly after treatment had started. A retrospective reanalysis of the first sample, using FISH, revealed that the trisomy 13 was present in a few divisions that had been completely unanalyzable by conventional cytogenetics. However, in general the chance of getting an informative result from these on-treatment samples is so low as to make them an unjustified waste of laboratory time.

If a clone is found to persist in the d 28 assessment bone marrow sample that is usually taken in acute leukemias, then it is an indication of a poor prognosis. The author's department has also found that up to 12% of patients in apparent cytologic remission of acute leukemia may have a few persisting clonal cells. The incidence is higher when there has been a persistent cytopenia or hypoplasia more than 6 mo after starting treatment. Despite this, conventional cytogenetic studies are generally an inefficient assay of remission status, and a FISH study is better if the abnormalities identified at diagnosis are suitable for detection by this technique. In general, cytogenetics, FISH, and quantitative polymerase chain reaction (PCR) assays for assessing the effectiveness of treatment or the levels of minimal residual disease (MRD) can be used only for cases where a clone has already been identified.

Note that the incidence of clonal cells is often independent of the number of blast cells present: A clone may sometimes be found when blasts are <5%, and yet be undetected when there are 90% or more. This is partly because blast cells are not always easy to identify and count precisely, and also because the number of blast cells in division (and so available for cytogenetic study) may be in a different proportion to the number of normal cells in division.

Unlike the situation in the acute leukemias, in chronic myeloid leukemia (CML) the Philadelphia translocation persists in almost all cells whether or not the patient has had any conventional treatment. However, secondary abnormalities may disappear once treatment starts, and reappear when the disease accelerates or the acute phase develops.

8. A Normal Result in Leukemias

Most centers find that 50–70% of AMLs have a detectable chromosome abnormality; in ALLs the detection rate is higher, being over 90% in a few centers. In myelodysplastic syndromes (MDS) and myeloproliferative disorders (MPD) the incidence of clones detected is usually <50%. It was suggested that all cases of acute leukemia may eventually be shown to have an abnormality ***(7,8)*** but in practice there will always be some cases in which an abnormality cannot be demonstrated by conventional cytogenetics. As described in Chapter 3 (**Subheadings 1.4.** and **4.3.**), and Chapter 5 (**Subheading 2.8.**), cryptic abnormalities can sometimes be shown to be present by FISH techniques in the absence of any obvious chromosome rearrangement.

Therefore the finding of only karyotypically normal cells does not mean that there is no malignant clone present: It may be that the abnormality occurred at too low a level to be detected, or that it was not present uniformly in the bone marrow and was missed during the biopsy; it may also be that the abnormality was too subtle to detect, even by an experienced cytogeneticist; or that there was a gene rearrangement that did not involve a chromosome rearrangement. Consequently, "normal" may be best defined as "clone not detected."

It has been estimated that in an adult about 40 thousand million new marrow cells are produced every hour, so only a very small proportion is being examined. The more cells examined, the more confident the cytogeneticist can be that there is not a clone present, but time and resource constraints limit the number of divisions that can be analyzed. In many laboratories it is the policy that 25 metaphases are analyzed whenever possible, unless the presence of a clone can be established with fewer.

Even if a clone is detected after analyzing just a few divisions, some more should be analyzed in case of clonal variation and even multiple clones. Although the clinical and prognostic significance of many common abnormalities is well established, the effects of the presence of secondary abnormalities are still being determined. In some instances these appear to have a detrimental effect on prognosis ***(9)*** and in others a beneficial effect ***(10)***.

There is evidence that some cell types prefer different kinds of culture conditions; the cytogeneticist needs to ensure that the appropriate conditions are used, to maximize the likelihood of getting divisions from the tumor cells.

In some cases karyotypically normal cells of good morphology are found on the same slide as karyotypically abnormal cells of poor morphology. This may give some reassurance that the poor morphology is not due to a technical failure, and is also a reminder against the risk of selecting only good-morphology divisions for analysis: It can be tempting to disregard the poor-morphology divisions and thereby exclude those that are abnormal.

However, it should never be assumed that good-morphology, karyotypically normal cells are not from the malignant clone; it may be that the primary clonal abnormality is invisible and the visible chromosome abnormalities are merely late-occurring events in the course of the disease.

9. A normal Result in Solid Tumors

Where cells have been cultured for some days or weeks, it is quite possible that normal fibroblasts will overgrow the malignant cells present. Even without such culturing, normal divisions may be predominant simply because the malignant cells are not in an active growth phase.

Malignant cells that have metastasised tend to be capable of vigorous growth. Consequently, divisions from a pleural effusion that is reactive to a benign condition will be normal, while those from a pleural effusion caused by metastasis are usually grossly abnormal.

10. Single-Cell Abnormalities

In many studies an occasional division is found with some chromosome abnormality. These tend to be more frequent after treatment, in patients who have been subject to some occupational or recreational exposure to clastogens, or who have a fragility syndrome, as mentioned in **Subheading 5.1.1.** However, they can occur in anyone.

If there are other divisions with single-cell abnormalities, and none of these is typical of the stated diagnosis, then it is usually safe to assume that they all have a nonclonal origin. It is the policy of many laboratories not to mention these single-cell abnormalities when reporting the study, to avoid creating any concern or confusion.

If a clone has not been already found, or the abnormality is one that is known to be recurrent in malignancy, then further studies should be undertaken to try to determine whether a single abnormal cell is from a clone or not. It is usually best to begin by analyzing or examining more divisions. If this does not provide a confirmatory result, then a different approach should be used, such as FISH or a molecular assay ***(11)***.

If these are not available, then a decision has to be made as to whether or not to inform the clinician about the finding. In the author's experience, most single-cell abnormalities are never seen again, but it has occasionally happened that a single-cell abnormality found in one study has recurred in subsequent studies, indicating that it was clonal. Therefore the author's policy is that if there was no clone found, then single-cell abnormalities are included in the report, with a statement warning that they may be of no significance.

References

1. ISCN (1995) *An International System for Human Cytogenetic Nomenclature.* (Mitelman, F., ed.), S. Karger, Basel.
2. Casalone, R., Granata Casalone, P., Minelli, E., et al. (1992) Significance of the clonal and sporadic chromosome abnormalities in non-neoplastic renal tissue. *Hum. Genet.* **90,** 71–78.
3. Brodeur, G. M., Dahl, G. V., Williams, D. L., Tipton, R. E., and Kalwinsky, D. K. (1980) Transient leukemoid reaction and trisomy 21 mosaicism in a phenotypically normal newborn. *Blood* **55,** 69–73.
4. Secker-Walker, L. M. and Fitchett, M. (1995) Commentary. Constitutional and acquired trisomy 8. *Leukemia Res.* **19,** 737–740.
5. United Kingdom Cancer Cytogenetics Group (UKCCG) (1992) Loss of the Y chromosome from normal and neoplastic bone marrows. *Genes Chromosomes Cancer* **5,** 83–88.
6. Heim, S. and Mitelman, F. (1995) *Cancer Cytogenetics,* 2nd edit. Alan R. Liss, New York.
7. Yunis J. J. (1981) New chromosome techniques in the study of human neoplasia. *Human Pathol.*, **12,** 540–549.
8. Williams, D. L., Raimondi, S., Rivera, G., George, S., Berard, C. W., and Murphy, S. B. (1985) Presence of clonal chromosome abnormalities in virtually all cases of acute lymphoblastic leukemia. *N. Engl. J. Med.* **313,** 640–641.
9. Hiorns, L. R., Swansbury, G. J., Mehta, J., et al. (1997) Additional abnormalities confer worse prognosis in acute promyelocytic leukemia. *Br. J. Haematol.* **96,** 314–321
10. Rege, K.., Swansbury, G. J., Atra, A. A., et al. (2000) Disease features in acute myeloid leukemia with t(8;21)(q22;q22). Influence of age, secondary karyotype abnormalities, CD19 status, and extramedullary leukemia on survival. *Leukemia and Lymphoma* **40,** 67–77.
11. Kasprzyk, A., Mehta, A. B., and Secker-Walker, L. M. (1995) Single-cell trisomy in hematologic malignancy. Random change or tip of the iceberg? *Cancer Genet. Cytogenet.* **85,** 37–42.

13

Cytogenetic Studies Using FISH

Background

Toon Min and John Swansbury

1. Introduction

Prior to the early 1970s, chromosome spreads were block stained with, for example, orcein or Fulgen's stains, and only those with a distinctive outline could be recognized. Then it was discovered that chromosomes could be made to show a consistent pattern of lighter or darker stained segments (bands) by using fluorescent dyes (fluorochromes) such as atebrin and quinecrine, or by treatment with agents such as trypsin, detergent, or a salt solution (e.g., saline sodium citrate), followed by staining with basic nuclear dyes such as Giemsa, Wright's, or Leishman's stain. Once every chromosome could be identified by its unique banding pattern, and recurrent abnormalities could be associated with specific diseases or physical disorders, the science of cytogenetics quickly proved to have direct and practical clinical applications. The chromosomes obtained in studies of malignancy, however, are often of poor morphology, and tend to be involved in complex and subtle rearrangements; in such cases, conventional cytogenetic studies are unable to define fully the entire karyotype. This limitation has been overcome by the

From: *Methods in Molecular Biology, vol. 220: Cancer Cytogenetics: Methods and Protocols*
Edited by: John Swansbury © Humana Press Inc., Totowa, NJ

introduction of new techniques, known as *in situ* hybridization (ISH), which bind labeled DNA to specific parts of the chromosomes being studied. Molecular techniques such as polymerase chain reaction (PCR) are used to test DNA that has been fragmented and amplified. However, ISH binds specially prepared probe DNA to the target DNA in chromosomes that are already spread on a slide.

Much of the early ISH work used probe DNA that had a radioactive label. This required special precautions against the risk of exposure to dangerous radioactivity. Its applications were limited, as only one DNA probe could be used at a time, and it was also a very slow process. Immunocytochemical reagents such as horseradish peroxidase can also be used to label DNA. This approach tends to be used by histopathologists rather than by cytogeneticists; its use is limited but it has the advantage of not requiring a fluorescence microscope.

The introduction of DNA probes labeled with fluorochromes <20 yr ago made the techniques safer, faster, and more flexible. However, it is only during the past 10 yr or so that (1) the cost of the equipment and reagents has fallen sufficiently to come within the reach of routine laboratory service, and (2) the reagents and technique have become standardized and reliable.

The acronym FISH designates fluorescence *in situ* hybridization. FISH is now widely accepted as a powerful adjunct to conventional cytogenetics, both in constitutional studies and in the genetic analysis of malignancy ***(1,2)***. Some FISH probes can also be used in interphase (nondividing) cells, which overcomes another restriction on conventional cytogenetic studies, the need for the contracted chromosomes that are produced during cell division. Valuable as FISH studies are, they do have their own limitations. Therefore, FISH is most effective as a complementary technique, rather than as a competitor with conventional cytogenetic analysis. The relationship between these techniques is explored further in Chapter 17.

For both research and clinical applications a great variety of DNA probes are now commercially available (e.g., from Oncor, Gaithesburg, USA, and Vysis, UK). Probes can be produced that identify whole chromosomes, parts of chromosomes, centromeres, telomeres, genes, and parts of genes. These probes can also be used in various combinations when investigating chromosome abnormalities.

1.1. Whole Chromosome Paints

Whole chromosome "paints" (WPCs) are composed of a mixture of many DNA probes that are specific to unique DNA sequences along the entire length of the target chromosome. Any DNA sequence that is duplicated on other chromosomes is suppressed, to avoid cross-hybridization. Partial chromosome paints similarly cover part of a chromosome, for example, the short arms of chromosome 3. These paints are prepared from flow sorted or microdissected, purified human chromosomes. Good quality paints are commercially available for the entire human karyotype. They provide uniform cover for almost all of each chromosome, being less effective at the centromere and telomeres (*see* **Fig. 1A**).

Chromosome abnormalities in malignancy range from simple balanced reciprocal translocations to complex rearrangements involving many chromosomes ***(3)***. The use of WCP can result in the full identification of all the chromosomes involved. Even apparently simple reciprocal translocations can sometimes be shown to be more complex by this technique, and sometimes morphologically normal chromosomes can be shown to have cryptic rearrangements ***(4,5)***.

The use of WPCs in interphase cells is very limited, as during interphase the chromosomes are dispersed in the cell nucleus and do not form discrete units.

A recent extension of the use of WCPs is multiplex FISH (MFISH), in which all 24 different chromosomes are simultaneously painted with a different color in a single hybridization experiment. The application of this technique is described in more detail in Chapter 16, and produces a result that is similar to spectral karyotyping (SKY) ***(6)***. The main difference between these two techniques is the way in which the colored images are captured. In MFISH a highly-sensitive monochrome camera captures a series of images that have passed through different filters, each allowing through light of

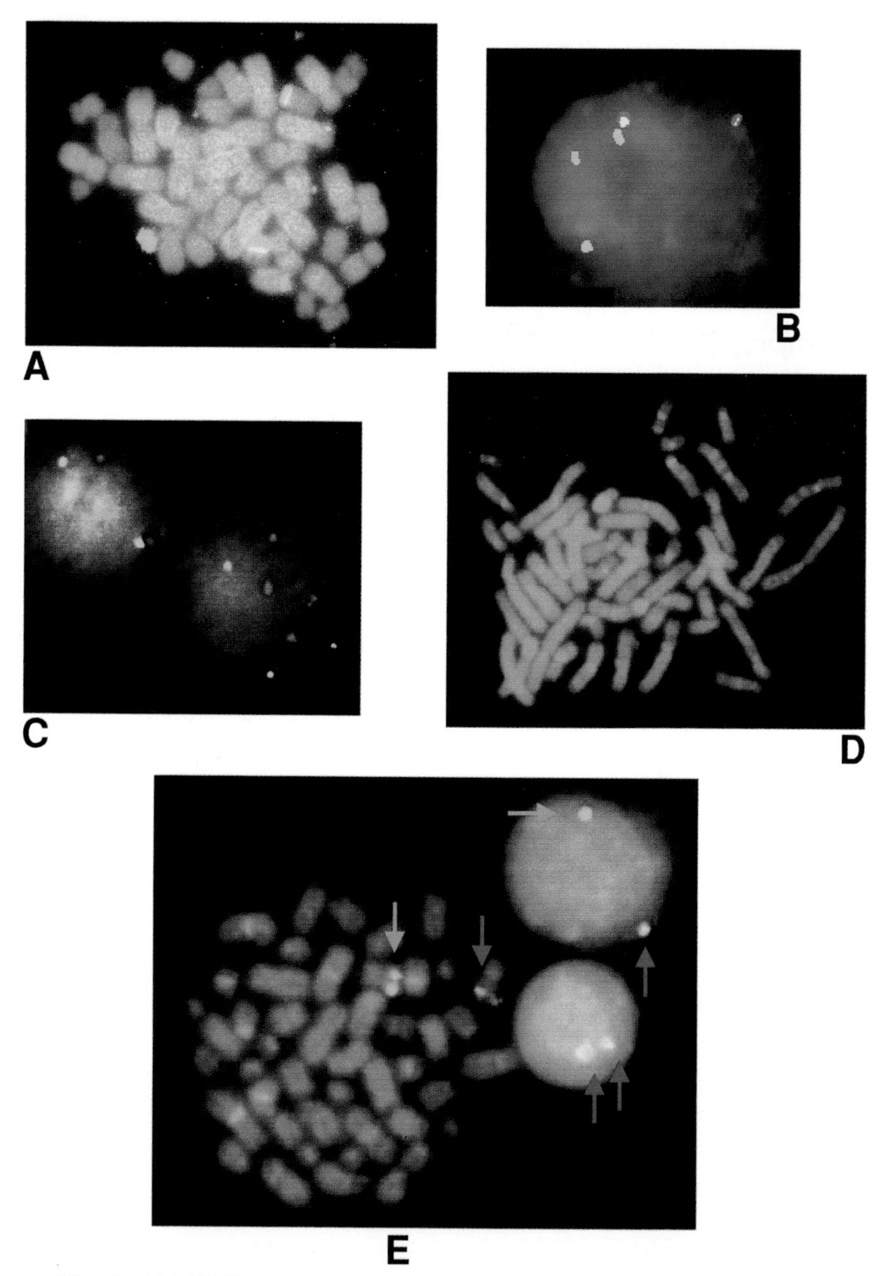

Fig. 1. (A) WCPs used to identify a t(8;21)(q22;q22). The normal 8 (*red*) and 21 (*green*) are on the left. Notice that the WCP for chromosome 21 cross-hybridizes with the centromeres of the two no. 13 chromosomes (indicated by *green arrows*). In this case, the paints provide uniform cover

specific wavelengths. These images are analyzed by a computer program that measures the contribution of each wavelength and calculates the ratio that is specifically associated with each chromosome. In SKY, the color spectrum for each pixel is obtained by splitting and recombining the light and measuring the resulting interference. Each technique has its strengths; in general, however, MFISH has greater flexibility for routine use.

1.2. Alpha Satellite (or Centromere) Probes

Centromere probes are derived from highly repetitive satellite human DNA sequences, which in most cases are located at highest concentrations in the centromeric regions of the chromosomes. These probes contain no interspersed repeat sequences derived from redundant DNA and the hybridization target is large, typically >1 Mb. Because the signal is large yet tightly localized, these probes can easily be seen in interphase nuclei as well as in metaphase chromosomes

---

for almost all of the chromosomes. However, WCPs are often less effective at the centromere and telomeres. **(B)** Centromeric (alpha satellite) probes for chromosome 8 (*red*) and chromosome 6 (*green*). This patient had a diagnosis of ALL, but all the divisions analyzed were normal. This FISH study showed trisomy 6 as part of a clone. **(C)** Locus-specific probes for *TEL/ETV6* (*green*) and *AML1* (*red*) (Vysis). These two cells are from the same patient as in **Fig. 2**. The cell on the *left* is normal, with two pairs of signals. The cell on the *right* has no fusions, and therefore there was no t(12;21)(p13;q22). However, there was one extra green signal and two extra red signals, showing that the clone had trisomy 12 and tetrasomy 21. **(D)** Locus-specific probe for *PDGF* at 16p13 (Oncor). This probe is manufactured to cover both sides of the breakpoint and so it splits when this locus is involved in a translocation or inversion. The single *red* signal at the top of the figure identifies the normal chromosome 16. Lower down there is an inverted chromosome 16, with a split *red* signal on each arm. **(E)** Dual-color probe for *MLL* (Vysis). This patient had a t(9;11)(p21–22;q23). The *red arrows* indicate the normal, intact *MLL* gene with the dual-color probe. The *green arrows* indicate the partly deleted *MLL* gene: the proximal, *green* signal is retained on the 11q, but the distal, *red* signal has been lost.

(*see* **Fig. 1B**). However, currently available centromeric probes for chromosomes 13 and 21 cross-hybridize, as do those for 14 and 22; it has not yet been possible to produce probes that are specific for each member of these pairs.

Centromeric probes are particularly helpful in identifying chromosomes that are dicentric. They rarely provide information about structural rearrangements involving other parts of the chromosome.

Aneuploidy, the gain or loss of whole chromosomes, is a relatively common finding in all kinds of malignancy, and can be effectively determined by using centromeric probes. High hyperdiploidy (>50 chromosomes), which occurs most frequently (up to 30% of cases) in childhood acute lymphoblastic leukemia (ALL), is associated with a good prognosis ***(7,8)***, so it is clinically important to identify patients in this group. FISH analysis using centromeric probes in these patients is particularly useful ***(9)*** because the chromosome morphology is often too poor for accurate analysis by conventional cytogenetic studies, yet some gains are associated with a better prognosis than others ***(10)***. In the United Kingdom, a centralized service, financed by the Leukemia Research Fund, screens by FISH all children with ALL in whom a cytogenetic study failed or found only normal divisions, primarily to identify those cases with good-risk high hyperdiploidy, and also the rarer cases with poor-risk near-haploidy. Individual numerical gains such as trisomy 8 and trisomy 21 are also common as secondary abnormalities in the acute leukemias ***(11)*** whereas trisomy 12 occurs predominantly in chronic lymphocytic leukemia (CLL) ***(12)***. Conventional cytogenetic studies in CLL are notoriously difficult (*see* Chapter 8), and many laboratories now routinely screen for trisomy 12 using FISH. The loss of chromosome 7 in acute myeloid leukemia (AML) indicates a poor prognosis and conventional cytogenetics may fail to detect this abnormality if the mitotic index is low. FISH is therefore useful in identifying monosomy 7, for which the detection rate has been shown to be underrepresented by conventional cytogenetics ***(13–16)***.

Although less widely used, also available are the beta satellite probe (which locates near to thecentromeres of 1, 9, and the acrocentric chromosomes), and satellite 1 DNA (which locates to the heterochromatic parts of chromosomes 1, 9, 16, and Y).

1.3. Locus- (or Gene-) Specific Probes

These are usually collections of one or a few cloned sequences homologous to restricted chromosome loci. Successful hybridization can be performed using probes as short as 1000 base pairs (=1 kb). Hybridization of any interspersed repeats in these probes is competitively inhibited by using protocols in which the signals derived from any ubiquitous sequences (i.e. present in other chromosomes) are suppressed (*see* **Figs. 1C,D**).

Commonly occurring translocations such as t(8;21)(q22;q22), t(15;17)(q24;q21), and inv(16)(p13q22) in AML, which have a favorable prognosis ***(17)***, and t(9;22)(q34;q11), which is associated with a poor prognosis ***(18–20)***, can now be readily identified with gene-specific probes. This is particularly helpful in cases in which these genes are suspected to be involved in complex translocations, or when the chromosome morphology is poor.

An example of a cryptic abnormality is t(12;21)(p12;q22), which is almost impossible to detect by conventional cytogenetics when it is the sole abnormality. It occurs in about 25% of cases of childhood pre-B ALL, having been identified by FISH using WCPs, gene-specific probes (for the *TEL* and *AML1* genes), and by other molecular genetic approaches such as Southern blotting and reverse transcription polymerase chain reaction (RT-PCR) ***(21–24)***. Furthermore, what was previously thought to be trisomy 21, seen in some cases of paediatric ALL other than as part of a high hyperdiploid clone, is now known in many cases to be a derivative of this translocation ***(25)***. Identifying this subtle abnormality is profoundly important in the clinical management of these patients because of its association with a good initial response but with a tendency to late relapse.

Another subtle abnormality is the translocation t(9;11)(p21–22;q23), usually associated with AML M5 ***(26–28)***, which can be difficult to see if the chromosome morphology is poor but that is readily identified with an *MLL* probe. The *MLL* gene (also known as *ALL1* and *HTRX*) is located at band 11q23 and has been found to be involved in translocations with >30 partner chromosomes. Abnormalities of 11q23 occur in >50% of infant acute leukemias, when they confer a very poor prognosis, and in 80% of patients with AML secondary to treatment with topoisomerase II inhibitors ***(29–31)***. Because the *MLL* gene is involved with so many chromosome partners, it is now the focus of many studies being carried out to elucidate its role and significance in the pathogenesis of leukemias ***(32)***.

Further examples of cryptic abnormalities revealed by gene-specific FISH probes are submicroscopic deletions associated with translocations ***(33)***.

Fewer gene-specific probes are available for studies of solid tumors than for the leukemias, a consequence of the relatively limited number of cytogenetic studies and the generally greater clone complexity; before probes can be produced, the relevant genes have to be located and identified. However, those probes that are available are proving to be very helpful in distinguishing between morphologically similar tumors.

FISH probes have also been used to identify the origin of the amplified DNA that constitutes homogeneously staining regions (HSRs) and double minutes (DMs), which occur in a variety of tumor cells. The most common genes amplified are c-*myc* (in leukemias and breast cancer), n-*myc* (in neuroblastoma), and *her-2-neu* (in breast cancer).

There are some rearrangements that are beyond the detection sensitivity of current FISH probes. For example, a tandem duplication (or self-fusion) of the *MLL* gene is common in cases of AML with trisomy 11 ***(34)***, and this may identify a subgroup of patients with poorer outcome ***(35)***. At present, detection of this abnormality is dependent on the use of other complementary molecular techniques such as Southern blotting and RT-PCR, which have greater sensitivity for detecting small rearrangements.

1.4. Telomeric Probes

Telomeric and subtelomeric probes are located at, or very close to, the telomeres of the chromosomes, that is, the very ends of the chromosome arms. These are useful because many WCPs do not cover the entire chromosome, but tend to leave the ends unstained. Although telomeric probes are not so widely used as other kinds of probes, they have provided new information, such as demonstrating that in some cases what appears to be a simple deletion is actually an unbalanced translocation ***(36)***.

1.5. Fiber FISH

The Fiber FISH technique involves the mechanical stretching of genomic DNA subsequent to cells being lysed. The resulting extended chromatin can provide highly extended linear DNA strands available for investigation by probes that are specific for parts of a gene. This technique has had little exploitation so far, but is potentially capable of revealing subtle intragenic rearrangements detectable down the microscope ***(37)***.

1.6. PRINS

Primed *in situ* hybridization (PRINS) is a technique similar to PCR except that instead of using gel electrophoresis to separate amplified DNA sequences, they are amplified after the chromosomes have been spread onto a slide ***(38)***. A single specific primer is used in a single-cycle PCR reaction to incorporate labeled dUTP/dNTP. The particular strengths of this technique are the intensity and clarity of the signal and the speed of the procedure, which can be completed within one hour.

1.7. FICTION

The acronym FICTION stands for fluorescence immunophenotyping and interphase cytogenetics as a tool for the investigation of neoplasia. The technique combines FISH analysis of genetic defects with other tests that identify cell constituents such as cell surface markers. An example of its use has been the identification of the cell lineage of leukemic cells with specific chromosome abnormalities ***(39)***.

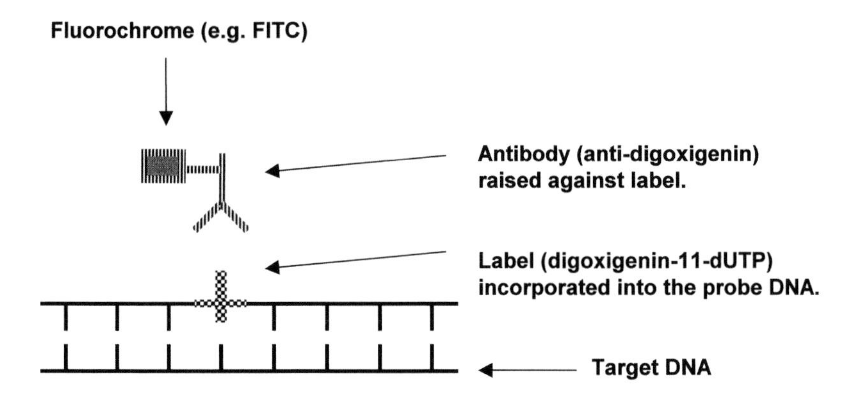

Fig. 2. Diagram to show a directly labeled probe, that is, one that has a fluorochrome previously attached to the probe DNA, allowing the probe DNA to be bound to the target DNA in one hybridization step.

2. Types of DNA Probes: Directly and Indirectly Labeled

Probes for FISH are grown in different vectors according to the size of the DNA fragment required. The vectors are plasmids (for probes of up to 10kb), phages (9–25 kb), cosmids (35–45 kb), or artificial chromosomes derived from bacteria (BACs, up to 300 kb), phage (PACs, 100–300 kb), or yeast (YACs, 200 kb–2 mb). To be able to visualize (microscopically) the hybridization of the probe to the target, a fluorescent label is attached to the probe, and probes are supplied as either directly or indirectly labeled. A directly labeled probe is one that has a fluorochrome previously attached to the DNA; this allows the probe DNA to be bound to the target DNA in one hybridization step (**Fig. 2**). An indirectly labeled probe has bound to it a hapten, which is a molecule such as biotin, digoxigenin, or estradiol, to which the fluorochromes can be attached. This is done after the probe has been hybridized to the target DNA. Using an indirectly labeled probe takes longer because of the extra stages of processing. However, an advantage of indirectly-labeled probes is that if a study is made using two or three probes, then the laboratory can choose which differently colored fluorochromes to use.

The fluorochromes most commonly used are fluorescein, rhodamine, coumarin, Texas Red, Spectrum Orange, and Spectrum

Green. In a busy hematology cytogenetics laboratory, the use of commercially produced, standardized, directly or indirectly labeled probes saves valuable time. However, very many unlabeled probes are also available that have been produced in a research context, and that have not yet been adopted for commercial development. These are often freely available, subject to certain restrictions, on request to the laboratory that produced them. They will require labeling by nick translation (an enzymatic labeling system that synthesizes nucleic acids using nucleotides that have a hapten). The nick translation procedure has been optimized to produce probe fragments of a length that are suitable for ISH (**Fig. 3**), and the technique is described in the next chapter. The most important parameter in the reaction is the action of DNase 1 which "nicks" double-stranded DNA in random locations, exposing free 3' OH groups. DNA polymerase 1 then adds nucleotides to these free 3' OH groups and simultaneously removes nucleotides from the 5' end. The nucleotides provided are hapten-labeled dUTP (e.g., with digoxigenin, biotin, or estradiol) and unlabeled dATP, dGTP, and dCTP. As the DNase 1 proceeds along the DNA, labeled dUTP is incorporated to produce the labeled probe. Insufficient nicking can lead to inadequate incorporation of the label and probes that are too long, whereas excessive nicking can produce probes that are too short.

3. Choice of Probes

Many commercial probes are now available, and these are generally easy to use, having been prepared to high standards of consistency and quality. Always read the marketing literature carefully, as there is a tendency not to provide adequate information about limitations. For example, some probes may not cover all the gene being tested, and some probes may not be contiguous, that is, there may be gaps in the length of DNA being covered, which in interphase nuclei can sometimes give the appearance of the signal being split.

In general dual-color probes are more informative, although more expensive. For example, the Vysis dual-color *MLL* probe located at 11q23 will detect the 25% of cases that have an unbalanced

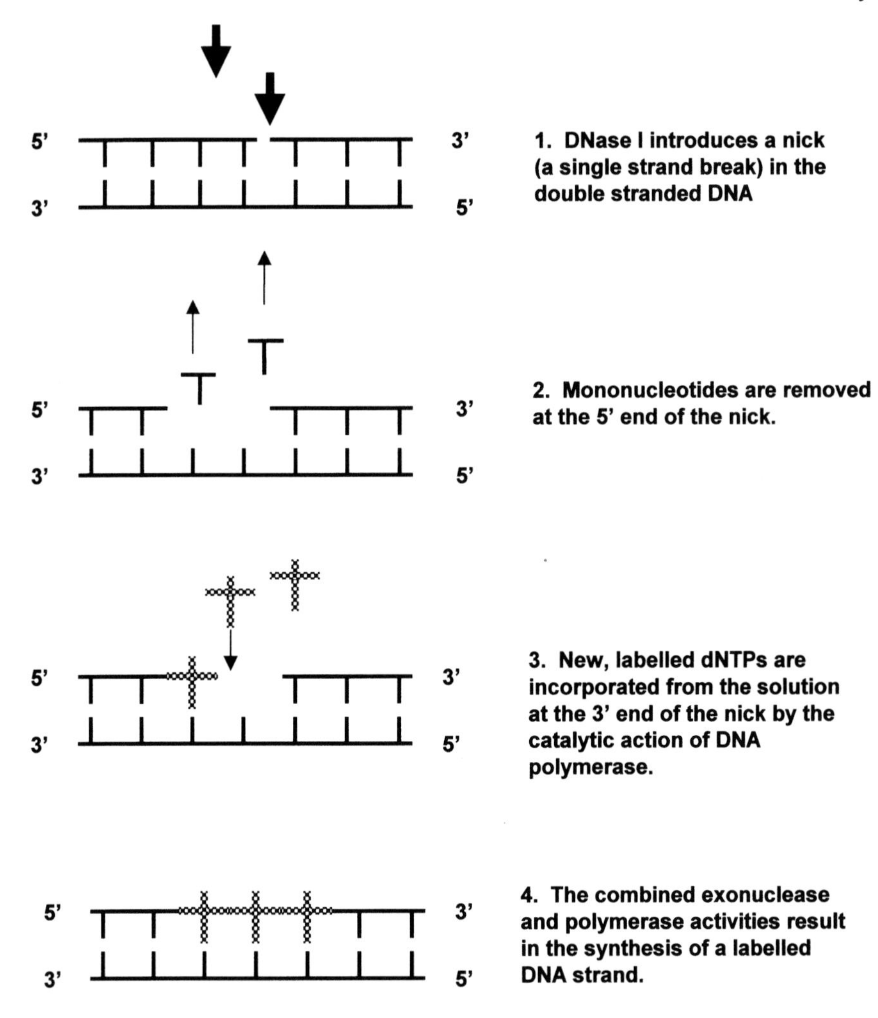

Fig. 3. Schematic representation of the incorporation of labeled deoxynucleotide triphosphates (dNTPs) into double-stranded DNA by nick translation.

translocation when part of the *MLL* gene has been deleted (*see* **Fig. 1E**). Some other *MLL* probes do not cover the whole gene, and so will not detect any translocations occurring outside the area covered.

In many cases of t(12;21)(p13;q22), the *TEL* gene on the remaining 12 is deleted, and this may correlate with a different response to treatment. Generally this deletion will be detected with the commercial probes currently available. However, in a few cases only a small part of the gene is deleted, and these cases may escape detection.

4. Analysis of FISH Preparations

Visualization and analysis of FISH signals from larger probes, such as chromosome paints, alpha satellite probes and YACs, can be successfully effected using a simple epifluorescence microscope with appropriate filter sets. Such a microscope has an ultraviolet (UV) light source as well as white light. The pathway for each kind of light is different, with white light passing through the slide and UV light being projected onto it. The detection of smaller signals (e.g., from phages) may require the help of computer-based image analysis systems that are now commercially available. A video camera or a low light charge-coupled device (CCD) camera is used to create a digital image, which is fed into a highly sophisticated, computer-based system that has dedicated image-analysis software as well as being able to control the filters, the camera, the exposure times, and the microscope. The combination of CCD cameras and image analysis makes it possible to process very faint signals and produce images with remarkable clarity. Two of the major manufacturers are Applied Imaging International, and Metasystems, GmbH, Germany, but other systems have also been developed by manufacturers including Leica Microsystems and Zeiss.

The slide is screened to locate suitable cells. Several images of each cell are usually collected, each through a different filter, with the computer controlling the exposure time for each filter. The computer analyzes these images, calculates the contribution made through each filter, and produces a pseudo-colored image based on the combined data.

When using FISH of interphase nuclei for assessment of response to treatment, for minimal residual disease, or for early detection of relapse, always bear in mind that the clone may have changed. Some of the chromosome abnormalities seen at diagnosis may have disappeared. For example, if a child with ALL had a near-haploid clone, and a FISH study using two or three centromeric probes was used to look for nuclei with missing chromosomes, it would fail to detect nuclei in which the entire near-haploid set had duplicated, which is a common development of this type of clone. Sometimes the entire clone is replaced by a new clone or an apparently new clone; this may mean that the patient has a new malignancy, or it may mean that the primary genetic abnormality was not visible.

5. Controls

Although commercially available probes are continually improving, and are produced to a high and consistent standard, the laboratory should always run its own controls before using the results from a new batch of probe. This will confirm that the correct probe has been supplied. In addition, batches of probe tend to vary in signal strength and in hybridization efficiency, and it may be necessary to establish new cut-off levels between background and positive signals. This is particularly important in the analysis of interphase nuclei, in which it is necessary to distinguish between true and false positives and negatives (*see* **Table 1**). A probe with a low hybridization efficiency will produce fewer signals, giving an underestimate of the number of targets present. This can result in a false-negative result, that is, a negative result from a sample that should have scored positive. Conversely, a probe with a high level of cross-hybridization and correspondingly high background may produce a false positive result.

As well as variation between batches of probe, there should be an assessment of variation between individuals who are performing the scoring ***(40)***. Useful guidelines on assessing the cutoff values and determining the sensitivity of FISH analysis have been described ***(41,42)***.

References

1. Bentz, M., Dohner, H., Cabot, G., and Lichter, P. (1994) Fluorescence *in situ* hybridisation in leukemia. The FISH are "spawning." *Leukemia* **8,** 1447–1452.
2. Berger, R. (1995) Recent advances in fluorescence in situ hybridisation (FISH) in hematology. *Pathol. Biol.* **43,** 175–180.

Table 1
True and False Positive and Negative Results

		Sample Being Tested	
		Positive/abnormal	Negative/normal
Result obtained by the analysis	Positive/abnormal	**True positive**	**False positive:** Result resembles that obtained in the presence of a clone, but no clone is actually present.
	Negative/normal	**False negative:** Failure to detect the presence of a clone	**True negative**

Diagram to illustrate the differences between true and false positives and negatives. If the sample being tested was abnormal but a negative or normal result was obtained, then this result was incorrect and was a false negative: it failed to detect the clone that was present. False positives are less common but can occur due, for example, to contamination, a positive result being obtained when the sample was actually normal.

3. Hiorns, L. R., Swansbury G. J., and Catovsky D. (1995) An eight-way variant t(15;17) in acute promyelocytic leukemia elucidated using fluorescence in situ hybridization. *Cancer Genet.Cytogenet.*, **83,** 136–139.
4. Saitoh, K., Miura, I., Ohshima, A., et al. (1997) Translocation t(8;12;21)(q22.1;q24.1;q22.1): a new masked type of t(8;21)(q22;q22) in a patient with acute myeloid leukemia. *Cancer Genet. Cytogenet.* **96,** 111–114.
5. Jadayel, D., Calabrese, G., Min, T., et al. (1995) Molecular cytogenetics of chronic myeloid leukemia with atypical t(6;9)(p23;q34) translocation. *Leukemia* **9,** 981–987.
6. Speicher, M. R., Ballard, S. G., and Ward, D. C. (1996). Karyotyping human chromosomes by combinatorial multi-fluor FISH. *Nat. Genet.* **12,** 368–375.
7. Secker-Walker, L. M., Prentice, H. G., Durrant, J., Richard, S., Hall, E., and Harrison, G. (1997) Cytogenetics adds independent prognostic information in adults with acute lymphoblastic leukemia on MRC trial UKALL XA. *Br. J. Haematol.* **96,** 601–610.
8. Pui, C-H., Rebeiro, R. C., Campana, D., et al. (1996) Prognostic factors in acute lymphoid and acute myeloid leukemias in infants. *Leukemia* **10,** 952–956.
9. Ritterbach, J., Hiddemann, W., Beck, J. D., et al. (1998) Detection of hyperdiploid karyotypes (> 50 chromosomes) in childhood acute lymphoblastic leukemia (ALL) using fluorescence in situ hybridization (FISH). *Leukemia* **12,** 427–433.
10. Mertens, F., Johansson, B., and Mitelman, F (1996) Dichotomy of hyperdiploid acute lymphoblastic leukemia on the basis of the distribution of gained chromosomes. *Cancer Genet. Cytogenet.* **92,** 8–10.
11. Zhao, L., Khan, Z., Hayes, K. J., and Glassman, A. B. (1998) Interphase fluorescence in situ hybridization analysis: A study using centromeric probes **7, 8,** and 12. *Ann. Clin. Lab. Sci.* **28,** 51–56.
12. Matutes, E. (1996) Trisomy 12 in chronic lymphocytic leukemia. *Leukemia Res.* **5,** 375–377.
13. Kolluri, R. V., Manueldis, L., Cremer, T., Sait, S., Gezer, S., and Raza, A. (1990) Detection of monosomy 7 in interphase cells of patients with myeloid disorders. *Am. J. Hematol.* **33,** 117–122.
14. Baurmann, H., Cherif, D., and Berger, R. (1993) Interphase cytogenetics by fluorescence in situ hybridization (FISH) for the characterization of monosomy-7-associated myeloid disorders. *Leukemia* **7,** 384–391.

15. Cotter, F. E. and Johnson, E. (1997) Chromosome 7 and hematological malignancies. *Hematology* **2,** 359–372.
16. Wyandt, H. E., Chinnappan, D., Ioannidou, S., Salama, M., and O'Hara, C. (1998) Fluorescence in situ hybridization to assess aneuploidy for chromosomes 7 and 8 in hematologic disorders. *Cancer Genet. Cytogenet.* **102,** 114–124.
17. Grimwade, D., Walker, H., Oliver, F., et al. (1998) The importance of diagnostic cytogenetics on outcome in AML: analysis of 1,612 patients entered into the MRC AML 10 trial. *Blood*, **92,** 2322–2333.
18. Tkachuk, D., Westbrook, C., Andreeff, M., et al. (1990) Detection of BCR-ABL fusion in chronic myelogenous leukemia by two-color fluorescence in situ hybridization. *Science* **250,** 559–562.
19 Werner, M., Ewig, M., Nasarek, A., et al. (1998) Value of fluorescence in situ hybridization for detecting the *bcr/abl* gene fusion in interphase cells of routine bone marrow specimens. *Diagn. Mol. Pathol.* **6,** 282–287.
20. Dohner, H. (1994) Detection of chimeric *BCR-ABL* genes on bone marrow samples and blood smears in chronic myeloid and acute lymphoblastic leukemia by in situ hybridization. *Blood* **83,** 1922–1928.
21. Romana, S. P., Mauchauffe, M., Le Coniat, M., et al. (1995) The t(12;21) of acute lymphoblastic leukemia results in *TEL-AML1* gene fusion. *Blood* **85,** 3662–3670.
22. Romana, S. P., Le Coniat, M., and Berger, R. (1994). t(12;21): A new recurrent translocation in acute lymphoblastic leukemia. *Genes Chromosomes Cancer* **9,** 186–191.
23. Shurtleff, S. A., Buijs, A., Behm, F. G., et al. (1995) TEL/AML1 fusion resulting from a cryptic t(12;21) is the most common genetic lesion in pediatric ALL and defines a subgroup of patients with an excellent prognosis *Leukemia* **9,** 1985–1989.
24. Wiemels, J. L. and Greaves, M. (1999) Structure and possible mechanisms of childhood acute lymphoblastic leukemia. *Cancer Res.* **59,** 4075–4082.
25. Loncarevic, I. F., Roitzheim, B., Ritterbach, J., et al. (1999) Trisomy 21 is a recurrent secondary aberration in childhood acute lymphoblastic leukemia with *TEL/AML1* gene fusion. *Genes Chromosomes Cancer* **24,** 272–277.
26. Sorenson, P. H. B., Chen, C-S., Smith, F. O., et al. (1994) Molecular rearrangements of the MLL genes are present in most cases of infant acute myeloid leukemia and are strongly correlated with monocytic or myelomonocytic phenotypes. *J. Clin Invest.* **93,** 429–437.

27. Chen, C-S., Sorenson, P. H. B., Domer, P. H., et al. (1993) Molecular rearrangements of 11q23 predominate in infant acute lymphoblastic leukemia and are associated with specific biological variables and poor outcome. *Blood* **81,** 2386–2393.
28. Swansbury, G. J., Slater, R., Bain, B. J., Moorman, A. V., and Secker-Walker L. M. (1998) Hematological malignancies with t(9;11)(p21–22;q23)—a laboratory and clinical study of 125 cases. *Leukemia* **12,** 792–800.
29. Heinonen, K., Mrozek, K., Lawrence, D., et al. (1998) Clinical characteristics of patients with de novo acute myeloid leukemia and Isolated trisomy 11: a Cancer and Leukemia Group B study. *Br. J. Haematol.* **101,** 513–520.
30. Pui, C-H., Behm, F. G., Raimondi, S. C., et al. (1989) Secondary acute myeloid leukemia in children treated for acute lymphoid leukemia. *New Engl. J. Med.*, **321,** 136–142.
31. Gill Super, H. J., McCabe, R., Thirman, M. J., et al. (1993) Rearrangement of the *MLL* gene in therapy-related acute myeloid leukemia in patients previously treated with agents targeting DNA-topoisomerase II. *Blood* **82,** 3705–3711.
32. Cimino, C., Rapanotti, M. C., Sprovieri, T., and Elia, L. (1998) *ALL1* gene alterations in acute leukemia: biological and clinical aspects. *Hematologica* **83,** 350–357.
33. Kolomietz, E., Al-Maghrabi, J., Brennan, S., et al. (2001) Primary chromosomal rearrangements of leukemia are frequently accompanied by extensive submicroscopic deletions and may lead to altered prognosis. *Blood* **97,** 3581–3588.
34. Caligiuri, M. A., Strout, M. P., Oberkircher, A. R., Yu, F., De La Chapelle, A., and Bloomfield, C. D. (1997). Partial tandem duplication of *ALL1* in acute myeloid leukemia with normal cytogenetics of trisomy 11 is restricted to one chromosome. *Proc. Natl. Acad. Sci. USA* **94,** 3899–3902.
35. Caligiuri, M. A., Strout, M. P., Lawrence, D., et al. (1998) Rearrangement of *ALL1* (*MLL*) in acute leukemia with normal cytogenetics. *Cancer Res.* **58,** 55–59.
36. Kearney, L. (2000) The impact of the new FISH technologies on the cytogenetics of hematological malignancies. *Br. J. Haematol.* **104,** 648–658.
37. Parra I., and Windle B. (1993) High-resolution visual mapping of stretched DNA by fluorescent hybridization. *Nat. Genet.* **5,** 17–21.

38. Pellestor, F., Girardet, A., Andreo, B., and Charlieu, J. (1994) A polymorphic alpha satellite sequence for human chromosome 13 detected by oligonucleotide primed in situ labelling (PRINS). *Hum. Genet.* **94,** 346–348.
39. Soenen, V., Fenaux, P., Flactif, M., et al. (1995) Combined immunophenotyping and in situ hybridization (FICTION)—a rapid method to study cell lineage involvement in myelodysplastic disorder. *Br. J. Haematol.* **90,** 701–706.
40. Dewald, G. W., Stallard, R., Alsaadi, A., et al. (2000) A multicenter investigation with D-FISH BCR/ABL1 probes. *Cancer Genet. Cytogenet.* **116,** 97–104.
41. Schad, C. R. and Dewald, G. W. (1995) Building a New Clinical Test for Fluorescence in situ hybridization. *Appl. Cytogenet.* **21,** 1–4.
42. Drach, J., Roka, S., Ackermann, J., Zojer, N., Schuster, R., and Fliegl, M. (1997). Fluorescence in situ hybridization: laboratory requirements and quality control. *Lab. Med.* **21,** 683–685.

14

FISH Techniques

Toon Min

1. Introduction

The impression is sometimes given that fluorescence *in situ* hybridization (FISH) studies are simply a matter of buying a kit with the right DNA probe, following the supplier's instructions, and reading a simple positive or negative result. In practice, getting a reliable result from a FISH study requires experience, time spent in testing to determine the precise local conditions needed for optimum hybridization, and time spent in assessing and scoring positive and negative controls to determine local baseline levels. The techniques described here will provide some useful guidelines about those aspects of the procedures that can be varied and those that are critical, and advice about the origin and resolution of commonly encountered problems. For simplicity, it will be assumed that the material to be studied is a fixed cell suspension, which is what is usually available in a cytogenetics laboratory. It is possible to study air-dried bone marrow or blood films, fresh tumor touch prints, and wax-embedded sections of solid tumors. The techniques are very similar to those described here ***(1,2)***. It is also possible to perform FISH studies on preparations that have already been banded and analyzed as part of a conventional cytogenetics study. However, the

From: *Methods in Molecular Biology, vol. 220: Cancer Cytogenetics: Methods and Protocols*
Edited by: John Swansbury © Humana Press Inc., Totowa, NJ

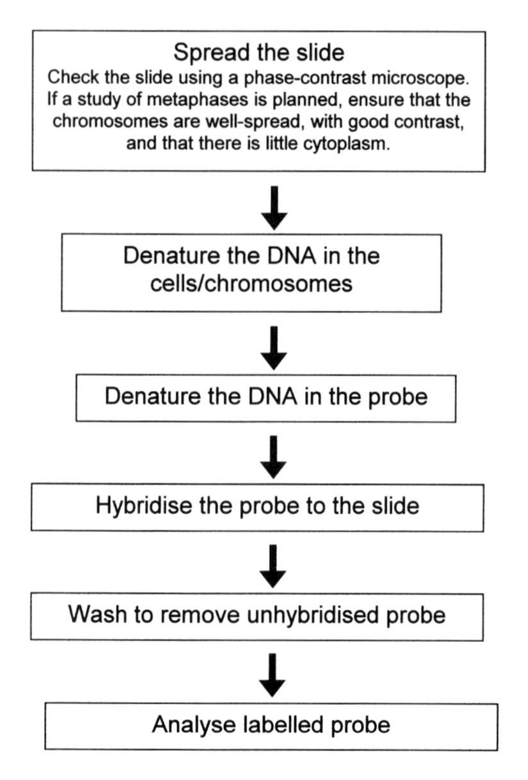

Fig. 1. Summary of FISH process.

simplest and most reliable methods use freshly spread cells, and this is what is described here.

A brief overview of the entire procedure is summarized in **Fig. 1**. The cells to be studied are harvested, fixed, and spread, as for conventional cytogenetic studies. The target DNA is either the metaphase chromosomes or dispersed in the interphase nucleus. It is treated with a formamide solution at a high temperature, which causes denaturation of the double strands of DNA; that is, they separate into single strands. Specially prepared, denatured, probe DNA is then added, which binds almost exclusively to parts of the target DNA that have the corresponding, matching sequence of nucleic acids. This probe DNA has been previously linked with a fluorescent dye, or with a hapten to which a fluorescent dye can be added later. The cells are also treated with a counterstain, that is, a general fluorescent dye that is usually either (1) 4',6-diamidino-2-phenylindole dihydrochloride (DAPI) which produces a faint G-banding pattern so that metaphase chromosomes can be at least partly identified as in a conventional cytogenetic study, or (2) propidium iodide. The slides are viewed under a microscope that has an ultraviolet light source. Fluorescence occurs when the electrons of a molecule of fluorochrome are excited by light of one wavelength and return to the unexcited state by emitting light of a longer wavelength. This fluorescence lasts for a limited time, after which it fades. The amount of light produced by fluorochromes is often very limited, especially if the probe is very small. Sometimes it cannot be seen clearly by the human eye. Therefore it is often necessary to photograph the cells, so that a record can be made before the fluorescence fades, and also so that the digitized image can be enhanced to make the colors brighter. This is usually done using a computer program specifically designed for cytogenetic studies.

2. Materials

Note: *Many of these reagents are harmful by contact, inhalation, and/or ingestion. It is important that principles of good laboratory practice are followed, and that appropriate health and safety precautions are taken.*

Most of these reagents can be obtained from any good supplier such as Sigma or GIBCO. Other suppliers are indicated where necessary.

2.1. Nick Translation

Most of the procedures use small quantities of reagents, and so are performed using 1.5-mL Eppendorf tubes. In some cases it is important to keep the reagents cold, so the tubes are inserted into crushed ice in a beaker or other container. Accurate measurement of very small volumes is needed; this can be done with a Gilson pipet.

The reagents are listed below in alphabetic order for ease of reference.

1. 50% Dextran sulfate: Warm a water bath to 65°C. Put 10 mL of 2× saline sodium citrate (SSC) into a flask and then add 10 g of dextran sulfate. Place in the water bath. Mix at intervals using the vortex mixer until the dextran sulfate has dissolved. Make up to 20 mL with more 2× SSC. Put 4-mL aliquots into tubes and store at –20°C.
2. 0.1M dithiothreitol (DTT): Prepare on ice in a fume cupboard. Dissolve 0.77 g of DTT in 5 mL of sterile H_2O, and store in 1-mL aliquots at –20°C.
3. DNase 1 (Boehringer Mannheim): 50 mg. Prepare on ice:

   1 mL of Tris-HCl, pH 7.4

   2.5 mL of NaCl

   0.5 mL of DTT

   5 mg of bovine serum albumin (BSA)

   25 mL of glycerol

   Make up to 50 mL with sterile H_2O, and store as 1-mL aliquots at –20°C. Immediately before use, dilute 1 in 1000 (i.e., 1 µL of 1 mg/mL of DNase1 in 1 mL of sterile H_2O).
4. 0.5 *M* EDTA: Dissolve 18.6 g of of disodium EDTA in 80 mL distilled water. Stir continuously with a magnetic stirrer. Add sodium hydroxide pellets until the pH reaches 8.0. Make up to 100 mL with more water. Autoclave before use.
5. Hybridization buffer:

   10 mL formamide (analytical grade)

   2 mL of 20× SSC

   4 mL of dextran sulfate

   2 mL of 10% Tween 20 (Pierce Laboratories, USA) (optional)

   2 mL of distilled water

   Mix well, dispense into 1-mL aliquots, and store at –20°C.
6. 1 *M* Magnesium chloride: Dissolve 101.6 g of $MgCl_2 \cdot 6H_2O$ in 500 mL of distilled water. Autoclave and store at room temperature.
7. 10x nick translation (NT) buffer

   5 mL of Tris-HCl, pH 7.4

   0.5 mL of 1 *M* $MgCl_2$

   5 mg of BSA

   Make up to 10 mL with sterile H_2O and store as 1-mL aliquots at –20°C.
8. dNTP nucleotides mix (Boehringer Mannheim): Prepare on ice:

   25 µL of 100 m*M* dATP

   25 µL of 100 m*M* dCTP

25 µL of 100 m*M* dGTP
25 µL of 100 m*M* dTTP

Make up to a 5-mL volume with sterile H_2O and store as 1-mL aliquots at –20°C.

9. 20× SSC: Dissolve 175.3 g of sodium chloride and 88.2 g of sodium citrate in 900 mL of distilled water. Adjust the pH to 7.0 using sodium hydroxide or hydrochloric acid, then make up to 1 L with more water. Store at room temperature for up to 6 mo.

   For other concentrations, either dilute this stock or else modify these amounts accordingly; for example, for 4× SSC use 34.08 g of sodium chloride and 17.64 g of sodium citrate per liter of solution.
10. Salmon sperm DNA, supplied at a concentration of 10 mg/mL. Store as 1-mL aliquots at –20°C.
11. 1 *M* Sodium chloride: Dissolve 58.4 g of NaCl in distilled water and make up to 1 L.
12. 3 *M* sodium acetate: Dissolve 40.8 g of sodium acetate·$3H_2O$ in 75 mL of distilled water. Adjust the pH to 5.5 with glacial acetic acid. Make up to 100 mL with more water.
13. TE buffer: Add 1 mL of 1 *M* Tris-HCl, pH 7.4 and 200 µL of 0.5 *M* EDTA and make up to 100 mL with distilled water.
14. 1 *M* Tris buffer: Dissolve 121.1 g of Tris base in 900 mL of distilled water. Use 1 *N* hydrochloric acid to bring the pH to 7.4 (which will need approx 65 mL), checking with a pH meter. (*Note:* not all pH meters have a probe that will cope with Tris; check before use.) Add more water to make up to 1 L of solution. Dispense into smaller bottles and autoclave before use. Store at 4°C.
15. Yeast RNA, final concentration 10 mg/mL. Prepare by dissolving in TE buffer. Store as 1-mL aliquots at –20°C.
16. The following solutions are used as supplied by the manufacturer:

    DNA polymerase
    Biotin-16-dUTP (Boehringer Mannheim)
    Digoxigenin-11-dUTP (Boehringer Mannheim)

2.2. Slide Denaturing and Post-Hybridization Washes

1. Ethanol series: 70%, 85%, and 100% absolute alcohol.
2. 70% Formamide in 2× SSC, pH 7: Mix 35 mL of formamide, 5 mL 20× SSC, and 10 mL of distilled water.
3. 1× SSC, pH 7: Dilute 10 mL of 20× SSC with 190 mL of distilled water.

4. 0.1× SSC, pH 7: Dilute 1 mL of 20× SSC with 199 mL of distilled water.
5. 2× SSC, pH 7: Dilute 10 mL of 20× SSC with 90 mL of distilled water.
6. 4× SSC, pH 7: Dilute 20 mL of 20× SSC with 80 mL of distilled water.
7. SSCT (4× SSC with 0.05% Tween-20) (Pierce Laboratories USA).
8. SSCTM (SSCT with 5% nonfat dried milk, e.g., Marvel).

2.3. Signal Detection

Note: Exposure to light causes fluorochromes to lose their ability to fluoresce. Therefore perform all procedures involving fluorochromes in reduced light. Any steps that do not require light, for example, incubations and washes, should be performed in darkness.

1. Fluorescein avidin DCS (Vector Laboratories Ltd.): Supplied as 1-mg protein concentration of 2 mg/mL Solution in 10 m*M* *N*-2-hydroxyethylpiperazine-*N*′-2-ethanesulfonic acid (HEPES) 0.5 *M* NaCl, pH 8.0. Keep in the dark at 4°C. Dilute 1:500 in SSCTM before use.
2. Texas Red avidin DCS (Vector Laboratories Ltd): Supplied as 2 mg/mL solution in 100 m*M* sodium bicarbonate, 0.5 *M* NaCl, pH 8.5. Keep in the dark at 4°C. Dilute 1:500 in SSCTM before use.
3. Biotinylated anti-avidin D (Vector Laboratories Ltd): Supplied as 0.5 mg active conjugate: Reconstitute in H_2O. The resulting solution will have the composition 10 m*M* HEPES, pH 7.5; 0.15 *N* NaCl. Store frozen in 10-µL aliquots. Dilute 1:100 in SSCTM before use.
4. Anti-digoxigenin–fluorescein and anti-digoxigenin–rhodamine, 200 µg of Fab fragments (Boehringer Mannheim): Dissolve in 1 mL of distilled H_2O then dilute 100 µL in 900 µL of distilled H_2O. Store in 1-mL aliquots at –20°C. Dilute 1:5 in SSCTM before use.
5. Counterstains: 10 µg/mL of DAPI or 0.5 mg/mL of propidium iodide (PI), dissolved in Citifluor anti-fade glycerol mountant (Citifluor Ltd., UK). This contains agents to reduce quenching (fading) of the fluorochromes, caused by oxidants or free radicals. The choice of counterstain is affected by the choice of fluorochrome being used. DAPI is better for red or green fluorochromes, such as Texas Red or rhodamine, and PI is better for yellow fluorochromes, such as fluorescein. The counterstain solutions and mountant can be bought separately and diluted as needed to final concentrations of 0.3 µg/mL of PI, or 0.1 µg/mL of DAPI.

2.4. Equipment

1. Very small volumes of liquids need to be dispensed accurately; a Gilson pipet is suitable, to which sterile disposable pipet tips can be attached.
2. A humid chamber is needed during hybridization. This can simply be a plastic box containing a slide rack and damp towels or tissues.
3. 1.5-mL Eppendorf tubes.
4. Two water baths.
5. Autoclave.
6. Vortex mixer.
7. Microfuge/microcentrifuge.
8. Ice machine, or access to crushed ice.
9. Incubator set at 37°C.
10. pH meter; this may need a special probe for testing Tris buffer.
11. Rubber solution for sealing the edges of coverslips. If this cannot be obtained from your usual supplier, it can often be obtained from shops that sell or repair bicycles.
12. Epifluorescence microscope (one that has an ultraviolet light source).
13. Fluorochromes: *See* **Table 1**.
14. Fluorescence filters: *See* **Note 1**.
15. A phase-contrast microscope is also very useful, for examining unstained slides to assess the quality of the spreading.

3. Methods

3.1. Probe Labeling by Nick Translation

As mentioned in the previous chapter, DNA probes labeled with haptens (i.e., biotin, digoxigenin, or estradiol) are commercially available (Oncor, Gaithesburg, USA, Vysis UK). DNA probes from noncommercial sources will usually require labeling by nick translation.

Before starting, prepare the solutions required, as described under **Subheading 2.1.**, and set two water baths to the temperatures that will be required, one at 16°C, and one at 68°C.

1. Stand an Eppendorf tube upright in a beaker of ice, and add the following reagents:

Table 1
A Selection of the Fluorochromes Available for FISH

	Excitation wavelength	Emission wavelength	Color of fluorescence
AMCA (aminomethyl coumarin acetic acid)	345 nm	440 nm	Blue
CY3	550 nm	570 nm	Red
CY5	650 nm	674 nm	Far-red
DAPI	360 nm	456 nm	Blue
FITC	495 nm	525 nm	Green
PI	342, 495 nm	639 nm	Orange-red
Rhodamine	550	575	Red
Spectrum Orange	545–555 nm	585 nm	Orange-red
Spectrum Green	505 nm	535 nm	Green
Spectrum Aqua	433 nm	480	Blue
TRITC (tetramethyl rhodamine isothiocyanate)	543 nm	570 nm	Red
Texas Red	596 nm	620 nm	Deep red

Probe DNA	n* µL
Biotin-16-dUTP or digoxigenin-11-dUTP**	2 µL
NT buffer	5 µL
DTT	5 µL
dNTP mix	4 µL
DNase 1	5 µL
DNA polymerase 1	2 µL

Make up to 50 µL with sterile distilled water.

*n is the volume of the probe solution that contains 1 µg of DNA

**Depending on the label required, add 2 µL of either digoxigenin-11-dUTP or biotin-16-dUTP to the dNTP mix.

2. Mix by flicking the tube, then pulse centrifuge for 2–5 s at 13,000 rpm.
3. Incubate for 2 h in a water bath at 15°C.
4. Replace the tube in the ice and add the following:

Human Cot-1 DNA (1 µg/mL)	100–200 µL
Yeast RNA (10 mg/mL)	5 µL
Sonicated salmon sperm DNA (10 mg/mL)	5 µL

Sodium acetate	5 µL
100% Absolute ethanol (ice-cold)	100 µL

5. Briefly vortex to mix.
6. Incubate at –20°C overnight or at –80°C for 1 h.
7. Centrifuge at 13,000 rpm for 10 min.
8. Discard the supernatant and leave the pellet to dry (30–60 min).
9. Resuspend the pellet in the following volume of hybridization buffer:
   For yeast artificial chromosomes (YACs), resuspend in 25 µL to give approx 250 ng/slide.
   For plasmid artificial chromosomes (PACs), resuspend in 50 µL to give approx 100 ng/slide.
   For cosmids, resuspend in 40 µL to give approx 40 ng/slide.
10. Store at –20°C. The probe should remain stable for some years.
11. The probe should be tested against positive and negative controls. If it does not give a good result, the probe fragment size may be too small or too large. This can be investigated using gel electrophoresis.

3.2. FISH Method

In many routine cytogenetics laboratories the techniques of probe labeling described in the preceding may be too time consuming. The following procedure describes the general method for FISH studies using commercially purchased probes. Always read the manufacturer's protocols carefully. The method described here is generally reliable but some probes may need specific alterations to the procedure.

The steps involved are listed below, and a detailed description follows:

1. Spread the slide.
2. Denature the target DNA on the slide.
3. Denature the probe.
4. Hybridize the probe to the target DNA.
5. Wash to remove any unhybridized probe.

3.2.1. Slide Preparation

For FISH studies of metaphases, standard culturing, harvesting, and fixation procedures are used, and these are described in detail in other chapters in this book. If a rapid result is needed and can be obtained from interphase cells, it is not necessary to wait to collect cells in division: simply suspend the cells in hypotonic solution for 10 min, fix, and then change the fixative three times.

Metaphase preparations can be made from both freshly fixed and archived fixed samples. In the author's laboratory, a large collection of fixed cytogenetic material going back 15 yr has been stored. Although there will be a degree of DNA degradation during this time, its quality is usually adequate for retrospective FISH studies. All that is necessary is to resuspend the fixed cells in fresh fixative for a short while, centrifuge, remove the supernatant, and then add a few drops of fresh fixative.

Slides for FISH studies are spread in the same way as those for cytogenetics studies (*see* Chapter 4). For high-quality FISH preparations, it is essential not to spread cells too densely on the slides, as this can increase background signal levels. Adequate fixative changes are also necessary to reduce cell debris, which can adversely interfere with a FISH analysis.

As the quality of the slides affects the formation of the metaphase spreads, it is necessary to use thoroughly clean, washed slides. Slides can be washed in ethanol and then kept in a freezer at –20°C before being used.

Drop the cell suspension (usually about 10-20 µL) on to a cold clean slide. Add three to four drops of fresh fixative on to the spread region and then leave to air-dry. It is helpful later if the area to be hybridized (usually about 22 × 22 mm) is defined by scoring underneath the slide with a diamond marker.

Check the slide using a phase-contrast microscope. If a study of metaphases is planned, ensure that the chromosomes are well spread, with good contrast, and that there is little cytoplasm. The chromosomes should appear dark gray, not black and shiny, or pale. Some cytoplasm can be cleared, if necessary, by using a pretreatment with RNase prior to hybridization, as described in **Subheading 3.2.1.1.**

Slides should be left to age overnight before being treated for FISH. However, if an urgent result is needed, then a satisfactory study can usually be made of slides just a few hours old. If slides are being prepared for FISH studies more than a few days later, they should be placed in a sealed slide box containing a silica gel desiccant, and stored at –20°C.

3.2.1.1. SLIDE PRETREATMENT. Freshly prepared slides do not usually require any enzyme pretreatment. However, metaphase spreads can be treated to facilitate disruption of the cell membrane and to allow efficient hybridization of the probe mixture to the target DNA. Three methods for slide pretreatment are given below; if the slides are over a month old or have been destained, then try method 1 or 2; if there is cytoplasm over the chromosomes, then try method 3.

Method 1. Incubate slides in 2× SSC for 30 min and rinse briefly in alcohol series. Air-dry.

Method 2. Rinse through an acetic acid series, diluted in water: 50%, 70%, and 100%. Rinse in an alcohol series (70%, 85%, and 100% absolute alcohol). Air-dry.

Method 3. Enzyme digestion with RNase:

RNase A, stock solution 10 mg/mL.

Add 10 µL of stock to 1 mL of 2× SSC to give 100 µg/mL in 2× SSC.

Place 100 µg on slide, add a coverslip, and incubate for 1 h in a humid chamber at 37°C.

Rinse briskly in two jars of 2× SSC at room temperature, for 3 min in each jar.

The response of the target cells to the above treatments varies according to the age of the slides. It may be necessary to experiment with different techniques and exposure times to obtain optimum results. Be warned that there is always a risk of losing the material from the slide, so test using a case with plenty of spare material.

3.2.1.2. PREHYBRIDIZATION. Probes are produced with the addition of Cot-1 human DNA as a blocking or competitor DNA that greatly improves signal clarity. This hybridizes to the sequences that are common to both the probe and the chromosomes being studied, thereby preventing hybridization of the probe DNA to these sequences. Only sequences specific to the target are available for probe hybridization *in situ*. If this were not done, then the end result would be multiple signals occurring on many chromosome sites as well as on the site of interest. Blocking DNA also hybridizes to molecules in the nucleoplasm and cytoplasm that could also bind to the probe. If a noncommercial probe is being used, then the amount of Cot-1 DNA added may have to be varied until all the unwanted sites have been blocked.

However, the blocking DNA may also prevent hybridization of probe DNA to closely adjacent DNA sequences by stearic hindrance. The method described here maximizes the rapid association of highly repeated sequences that are common to the probe and to the blocking DNA.

3.3. Slide Denaturation

The target DNA on the slides is denatured, that is, the DNA is rendered single-stranded to allow hybridization with the fluorescent-labeled probe. Ordinarily, DNA needs prolonged exposure to temperatures of >90°C to denature. However, using formamide, an organic solvent, allows denaturation to take place at lower temperatures.

Two methods for denaturation are described here. The second technique works successfully with most commercial and noncommercial probes, and is now routinely used in the author's laboratory. This method has the added safety feature of using less formamide, as well as reducing cost and waste.

3.3.1. Coplin Jar Method

1. Fill a Coplin jar with 70% formamide–2× SSC, and place in a water bath to heat up to exactly 70°C if one or two slides are being processed, 71°C for three slides, or 72°C for four slides. No more than four slides should be processed at any one time. Measure the temperature inside the Coplin jar, as this may be different from that of the water bath.
2. Prepare a Coplin jar with 70% ethanol and make ice-cold.

3. Prepare three Coplin jars with an ethanol series (70%, 85%, and 100%).
4. Immerse the slide(s) in the Coplin jar with formamide for 2 min.
5. Remove the slide(s) with forceps and arrest the denaturation by immersing in the ice-cold ethanol, and then in the ethanol series, giving 2 min in each jar. The slides may be left in the 100% ethanol until they are air-dried prior to the next step.
6. Air-dry the slide(s).

3.3.2. Hotplate Method

1. Prepare a Coplin jar with 70% ethanol and make ice-cold.
2. Prepare three Coplin jars with an ethanol series (70%, 85%, and 100%).
3. Warm the hotplate to 72° ± 1°C.
4. Place 80 μL of 70% formamide–2× SSC onto each slide.
5. Place a coverslip over the hybridization area.
6. Place the slide(s) on a hotplate for 1.5 min (it takes approx 30 s for the surface of the slide to reach 72°C).
7. Shake off the coverslip and dispose in a safe sharps container.
8. Immerse the slide(s) in the ice-cold ethanol and then in the ethanol series, giving 2 min in each jar. The slides may be left in the 100% ethanol until they are air-dried prior to the next step.
9. Air-dry the slide(s).

3.4. Probe Denaturation

Prepare a container with some crushed ice.

1. Aliquot the required amount of probe mixture (approx 10 μL/slide) into a clean Eppendorf tube.
2. Microfuge briefly
3. Denature in 70°C water bath for 10 min.
4. Plunge the tube into the ice.
5. Preanneal by placing in a 37°C incubator for 30–60 min.

3.5. Hybridization

This procedure should be conducted in a room with reduced light when working with directly labeled probes.

Place a humidified chamber into an incubator at 37°C.

1. Briefly centrifuge the probe, then preanneal as described in **Subheading 3.4., step 5**.
2. Warm the slide to 37°C.
3. Usually 10 µL of probe is enough for half a slide. Pipet this onto the slide and immediately place a coverslip over the area. Be careful to avoid the formation of air bubbles, as the probe will not hybridize uniformly around a bubble.
4. Seal around the edge of the coverslip with rubber cement and incubate the slide at 37°C overnight in a humid chamber.

3.6. Post-Hybridization Washes and Signal Detection

After hybridization is complete, unbound probe is removed by a series of washes. These washes are usually carried out in a slightly more stringent solution than the hybridization buffer, to denature and remove weakly bound probe (*see* **Note 2**). This should leave only the positively bound probe-target DNA. If there is a lot of background (unwanted hybridization to contaminating DNA) then the stringency can be increased by using 65% formamide instead of 50% in the hybridization buffer.

It is important that the slides are prevented from drying out at any stage before counterstaining and mounting.

1. Warm the wash solutions as follows:
   Place three jars with 1× SSC in a water bath and heat to 45°C.
   Place three jars with 0.1× SSC in a water bath and heat to 60°C.
2. Remove the rubber sealant and shake off the coverslip. The coverslip must be removed gently from the slide to avoid damage to the cells.
3. Wash three times in 1× SSC at 45°C for 5 min each.
4. Wash three times in 0.1× SSC at 60°C for 5 min each.
5. Wash in SSCT at ambient temperature for 2 min.
6. If using a directly labeled probe, drain off the SSCT and apply counterstain as described in **Subheading 3.7.**

If using an indirectly labeled probe, add the following steps:

7. Wash in SSCT at room temperature for 2 min.
8. Add 80 µL of SSTM and apply a 22 × 50 mm coverslip.

9. Incubate in humid chamber at 37° C for 20 min.
10. Remove coverslip by shaking off.
11. Immerse slide in SSCT for 2 min.
12. Remove excess fluid and add 80 μL of antibody-conjugated fluorochrome (e.g., avidin–Texas Red or anti-digoxigenin–FITC) and place coverslip (22 × 50 mm).
13. Incubate in humid chamber at 37°C for 20 min.
14. Remove coverslip and immerse slide in SSCT for 2 min. At this step a second antibody step can be applied for signal amplification; for example, for avidin–Texas Red, add biotinylated anti-avidin followed by a second round of avidin–Texas Red. Small probes such as cosmids may be better visualized with amplification. *See* **Note 3**.

3.7. Counterstaining

Slides should be mounted in Citifluor antifade mountant containing counterstain (e.g., PI or DAPI) by depositing one drop of about 20 μL on each slide. Apply a 22 × 50 mm coverslip. Carefully blot each slide to remove excess mountant. These slides may be stored for up to 6 mo if kept at 4°C in the dark.

3.8. Assessing the Result

It can take a while for eyes to adjust to seeing fluorescence; it is prudent to examine systematically several fields if the first field examined does not appear to have any signals. However, the FISH procedure is not infallible and sometimes fails to give a result even when performed in an experienced laboratory. The most likely causes of failure are listed in **Note 4**.

3.9. Screening and Analysis

The light emitted by fluorochromes is often very low, and it will be easier to see if the microscope is in a dark room or is surrounded by dark curtains.

If the probe is large, the signals are strong, and the hybridization has been efficient, then it is usually possible to screen a slide by eye and score the number of signals in each nucleus or metaphase. If the signals are small, or faint, or if the hybridization has been poor, then it can be difficult to see the signals by eye, and it may be necessary to capture (photograph) each nucleus or metaphase and then use a computer to enhance the image.

The hybridization efficiency can vary across a slide, so choose an area with well spaced cells, low background, and clear signals. Systematically work across the area and record how many nuclei have none, one, two, three, or more signals. If a dual-color probe is being used, record how many cells have split signals. Enough cells should be scored to give a clear and unambiguous result. For example, if a patient with a possible diagnosis of chronic myeloid is being screened and the first 20 cells all show a *BCR/ABL* fusion, then it is not necessary to do more to confirm the diagnosis. However, screening more cells might help to detect the presence of an extra der(22), a common secondary abnormality that has clinical significance. Conversely, if the same patient is being studied after a bone marrow transplant, when low levels of positivity would be expected, then several hundred cells may need to be screened.

Records should be kept of the hybridization efficiency of all the probes kept in stock. A probe with a reduced hybridization efficiency may fail to provide signals for all the genes in a cell, giving an underestimate of the true incidence. A control study (*see* **Note 5**) will indicate the expected distribution of one, two, and three or more signals. When the mean number of normal/abnormal control signals has been determined, a range of ± 3 standard deviations can be calculated. Any test results should be outside this range before they can be accepted as significant. This is particularly important when screening using a single-color probe.

A further complication is that some cells with apparently just one signal may actually have two signals that happen to be on top of each other. Similarly, a cell with three signals might be a tetraploid cell in which two of the four expected signals are again superimposed. These coincidences are likely to occur at only a low frequency. If they are likely to complicate the interpretation of the study (e.g., when screening for low levels of trisomy 8), then it would be wise to perform the study with two probes, one for the abnormality being investigated and one for another chromosome. Similarly, if screening for a small deletion (e.g., of the retinoblastoma gene at 13q14), two probes should be used, one for the rb1 gene and one located also on chromosome 13 but at some distance away, to distinguish between gene deletion and loss of the entire chromosome.

4. Notes

1. A filter is needed for each fluorochrome being used, for example, DAPI, fluorescein isothiocyanate (FITC), rhodamine. A dual-bandpass filter block is also recommended, for example, for FITC and Texas Red.

   The bandwidth of a filter is the range of wavelengths at which at least 50% of the light is transmitted. A broad-band filter allows more light to reach the specimen, so the signals appear brighter, but tend to fade more quickly. Narrow band pass filters will produce fainter signals, but less fading and less extraneous fluorescence.

   Most probes can be adequately identified using the broad-band Pinkel filter sets—for UV 405 nm, blue (490 nm), and yellow (570 nm). If more exotic fluorochromes are used then the appropriate specific filter sets for the emission and excitation wavelengths will be needed.

   Viewing the slide can be done through either single-band pass filters whereby only one color can be visualized at a time, or by using either a dual- or triple-bandpass filter when more than one color can be visualized simultaneously.

   Although the advantage of using multibandpass filters is that more than one color can be visualized on the slide at one time, the signals emitted by the probes are less bright than in single-bandpass filters. This is because each filter reduces the amount of light falling onto the specimen.
2. The stringency of post-hybridization washes is a description of how severe the washing process is. It has to be a compromise between insufficient washing, which will leave a high level of background "noise" due to probe being attached to DNA outside the area of interest, and excessive washing, which will give cleaner but fainter signals. The stringency of washes is affected by temperature as well as by the composition of the mixture. In general, high stringency is associated with high temperature and low salt concentration, while low stringency is associated with low temperature and high salt concentration.

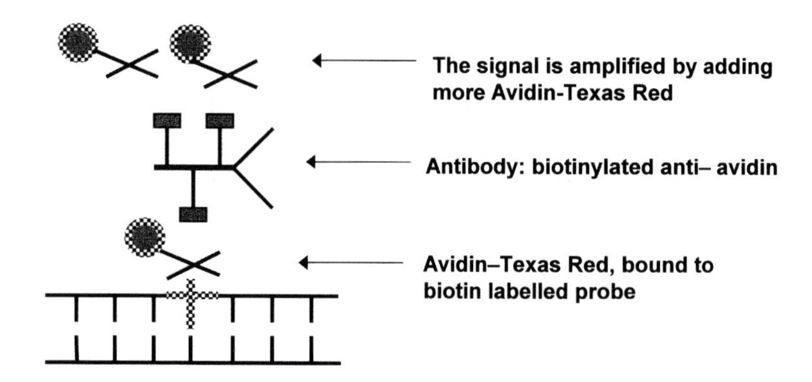

Fig. 2. Detection of biotin-labeled probe using an amplified biotin–avidin Texas Red system.

3. If the target sequence is large (> 50 kb), the signal is usually readily visible. Generally, whole chromosome probes, probes for repetitive sequences, and large probes cloned in artificial chromosomes can all be detected without amplification. If, however, the signal is not bright then it may be amplified by immunocytochemistry. Most probes are immunogenic and can be detected by immunocytochemical methods. For example, if a probe was labeled with avidin, then to this would be added a layer of biotinylated anti-avidin antibody, which provides additional binding sites for another layer of fluorochrome-labeled avidin (*see* **Fig. 2**). Since amplification is rarely needed with modern probes, a detailed protocol is not provided here.
4. Troubleshooting: Commercial DNA probes for FISH studies are usually supplied with instructions about how to prepare them for use. The impression is sometimes given that a result is guaranteed if the instructions are followed, and indeed this is often the case. However, problems do arise more often than the inexperienced might expect.

   The most commonly encountered problems are high background and poor signal strength. There can be many reasons for this ranging from incorrect dilutions of the fluorochromes to inadequate post-hybridization washing procedures. Lack of any signal may be due to incorrect probe concentration to labeling problems. An excellent guide to troubleshooting is contained in the Oncor FISH manual (Oncor, Gaithesburg) *(3)*. In addition, most commercial probe kits usually contain troubleshooting guidelines.

   The most common causes of problems are:

a. Faulty or incorrect probe was supplied.
b. The probe has become contaminated.
c. The probe has degraded (deteriorated) with age.
d. Faults with the technical procedure, usually due to incorrect preparation of solutions, inaccurate temperature control, or omission of a stage during processing.
e. faults with the analysis system, due to incorrect filters, aging of the UV light source, incorrect alignment of bulb, or quenching of the fluorochrome caused by too long exposure to light.

Whenever a new FISH study is performed, a parallel study must be made using cells from a case known to be positive for the abnormality being tested, and preferably also a third slide with cells from a case known to be negative. As well as acting as positive and negative controls against which to compare the test case, these will also give an indication of the hybridization efficiency of the probe and the amount of background.

5. Controls: Although commercially available probes are continually improving, it is always prudent to run tests using simple positive procedural controls such as alpha satellite probes or whole chromosome paints to metaphase chromosomes. This helps in identifying weak aspects of the laboratory's FISH procedure, faulty solutions, and so forth. In addition, when using any new probe it is advisable to use positive and negative controls to estimate the appropriate cutoff levels and sensitivity. This is particularly important in the analysis of interphase nuclei, in which it is necessary to distinguish between true and false results, both positive and negative (*see* Chapter 13). In any diagnostic service it is necessary to determine the local scoring criteria, and to define acceptable range of expected percentages for normal and abnormal results. Discrepancies of scoring between observers in the participating laboratories have been evaluated ***(4)***. Useful guidelines on assessing the cutoff values and determining the sensitivity of FISH analysis have been outlined by Schad and Dewald ***(5)*** and by Drach et al. ***(6)***.

Acknowledgments

I wish to thank John Swansbury for his encouragement and patience, Dr. Lionel Coignet and Shayne Atkinson for clarification of some aspects of the techniques, and Tracy Root for her help with the diagrams. I am also grateful to the Royal Marsden NHS Trust under whose auspices this chapter was written. This chapter is dedicated to Shenpen Dawa Rinpoche and A. S. B.

References

1. Dohner, H. (1994) Detection of chimeric *BCR-ABL* genes on bone marrow samples and blood smears in chronic myeloid and acute lymphoblastic leukemia by in situ hybridization. *Blood* **83,** 1922–1928.
2. Muhlmann, J., Thaler, J., Hilbe, W., et al. (1998) Fluorescence in situ hybridization (FISH) on peripheral blood smears for monitoring Philadelphia chromosome-positive chronic myeloid leukemia (CML) during interferon treatment: a new strategy for remission assessment. *Genes Chromosomes Cancer* **21,** 90–100.
3. *The Ultimate FISHing Guide: Sample Preparation and Application Protocols* (1996) Oncor, Gaithesburg.
4. Dewald, G. W., Stallard, R., Alsaadi, A., et al. (2000) A multicenter investigation with D-FISH BCR/ABL1 probes. *Cancer Genet. Cytogenet.* **116,** 97–104.
5. Schad, C. R. and Dewald, G. W. (1995) Building a new clinical test for fluorescence in situ hybridization. *Appl. Cytogenet.* **21,** 1–4.
6. Drach, J., Roka, S., Ackermann, J., Zojer, N., Schuster, R., and Fliegl, M. (1997). Fluorescence in situ hybridization: laboratory requirements and quality control. *Lab. Med.* **21,** 683–685.

15

FISH, CGH, and SKY in the Diagnosis of Childhood Acute Lymphoblastic Leukemia

Susan Mathew and Susana C. Raimondi

1. Introduction

Although classical cytogenetic analysis is a powerful tool for the assessment of acquired chromosomal changes in hematological malignancies, it can be performed only on dividing cells and cannot detect cryptic rearrangements. The introduction of molecular cytogenetic techniques, such as fluorescence *in situ* hybridization (FISH), has revolutionized the field of cytogenetics by allowing the identification of complex and cryptic chromosomal abnormalities. FISH allows the study of chromosome exchanges and gene rearrangements, amplifications, and deletions at the single-cell level. However, another FISH-based technique, comparative genomic hybridization (CGH), can identify chromosome losses and gains in tumor cells without prior knowledge of the chromosomal loci involved *(1)*. Furthermore, the capacity to hybridize simultaneously 24 or more DNA probes in the FISH-based karyotyping of chromosomes has resulted in several novel techniques, such as multiplex FISH (MFISH) *(2)*, spectral karyotyping (SKY) *(3)*, combined binary ratio labeling (COBRA) *(4)*, and color-changing karyotyping *(5)*. FISH banding techniques have also been developed to identify

From: *Methods in Molecular Biology, vol. 220: Cancer Cytogenetics: Methods and Protocols*
Edited by: John Swansbury © Humana Press Inc., Totowa, NJ

intrachromosomal rearrangements: cross-species color banding (CSCB) *(6)* and high-resolution multicolor banding *(7)*. FISH has numerous applications in the diagnosis and management of neoplastic disorders, particularly hematological malignancies. This chapter will focuses on the FISH, CGH, and SKY methods used in childhood acute lymphoblastic leukemia (ALL).

FISH techniques allow the detection of specific nucleic acid sequences (DNA or RNA) in metaphase chromosomes, interphase cells, or frozen tissue sections. In combination with immunocytochemistry, *in situ* hybridization (ISH) relates topographic information to gene activity at the DNA and mRNA levels. This chapter addresses FISH using DNA probes only.

1.1. Labeling of Probes

Probes for ISH procedures can be labeled by a variety of methods such as nick translation and random priming. The most widely used method is nick translation *(8)*. Two types of labeling, direct and indirect, are currently in use. In the direct method, the probe DNA is tagged with a fluorescent dye (e.g., Spectrum Orange, Spectrum Green, cyanine dyes) so that the probe–target complexes can be visualized immediately after their hybridization. Probes for indirect labeling methods have been chemically or enzymatically modified to carry a reporter molecule or hapten; the probe:target complexes are visible only after affinity cytochemical treatment. Biotin and digoxigenin are widely used as reporter molecules in probes that are indirectly labeled. Directly and indirectly labeled probes for centromeres, whole chromosomes, and a few unique DNA sequences are commercially available (Vysis, Downers Grove, IL; Cambio, Cambridge, UK; Cytocell Ltd., Banbury, UK; American Technologies, Arlington, VA). It should be noted that the probes that are made commercially vary in design, and companies change from time to time.

1.2. Types of probes

A variety of probes, each having a different cytogenetic application, are available for use in FISH.

1. Repetitive sequence probes target specific regions of chromosomes. Satellite repetitive probes target DNA sequences that are tandemly repeated several hundred times in the centromeric (alpha satellite probes) and heterochromatic (beta and classical satellite) regions of chromosomes. These repeat sequence probes are used to detect numerical chromosomal changes (aneuploidy) but cannot identify structural abnormalities.
2. Unique sequence/single-copy probes contain DNA homologous to specific human genes, loci, or regions. These probes are cloned in cosmid, plasmid, phage, yeast artificial chromosome (YAC), plasmid artificial chromosome (P1), PAC, and bacterial artificial chromosome (BAC) vectors. They detect micro- deletions, amplifications, and rearrangements (translocations) in interphase nuclei and metaphases. Inversions can also be detected in metaphases by using these probes. The high specificity and efficiency of these probes help to ascertain structural chromosomal aberrations at the one-cell level.
3. Whole chromosome painting probes or arm-specific sequence probes contain a complete set of DNA sequences from one chromosome or from one chromosome arm. These probes are derived from flow-sorted chromosomes, chromosome-specific libraries, or microdissected DNA specific for each chromosome or chromosome arm ***(9–11)***. The probes are chromosome-specific, contain both repetitive sequences and unique sequences of each chromosome, and are used for the identification of chromosomal sequences involved in translocations, marker chromosomes, and rings. However, FISH with painting probes is limited to metaphase spreads.
4. Subtelomeric probes contain a locus estimated to be within 300 kb of the end of the chromosome that contains unique sequences and is specific for a single human chromosome arm ***(12)***. These subtelomeric regions represent a major diagnostic challenge in clinical cytogenetics, because most of the terminal bands are G-negative and have limited banding resolution. FISH analysis with subtelomeric probes can detect cryptic deletions and rearrangements of these regions that are not detected by conventional cytogenetics ***(13–14)***. However, because of the polymorphism of these variants, only 2.6% of the findings may have any clinical significance ***(13)***.
5. SKY probes are a combination of 24 differentially labeled probes that are generated from chromosome-specific DNA libraries obtained by flow sorting human chromosomes. These probes are amplified by using a degenerate oligonucleotide primer polymerase

chain reaction (DOP-PCR) ***(15)*** and are labeled by using 5 different fluorochromes and their combinations to achieve 24 colors. The distinction between the dyes can be attained only with the SD-200 Spectracube™ spectral imaging from Applied Spectral Imaging, Ltd. (ASI), Migdal Haemek, Israel.

6. MFISH probes represent 22 autosomes and two sex chromosomes that are combinatorially labeled by nick translation with fluorescein isothiocyanate (FITC), CY3, Cy3.5, Cy5, or Cy7. These probes are generated by microdissection and amplified by PCR *(2)*.
7. Cross-species banding probes consist of chromosome-specific probes that are generated by DOP-PCR directly from flow-sorted chromosomes from gibbons. Using combinatorial labeling of these probes and hybridizing on human metaphases can delineate the entire human karyotype in many painted segments in a multicolor format referred to as cross-species color segmenting or banding ***(6)***.

2. Materials

1. Incubator: set at 37°C.
2. Microcentrifuge.
3. Humidified chamber.
4. Water baths: 37°C, 43–45°C, 70°C.
5. Rubber cement.
6. Precleaned microscope slides and coverslips.
7. Coplin jars.
8. 0.5-mL Polypropylene microcentrifuge tubes.
9. Slide warmer (optional).
10. Microliter pipetting devices (P1000, P200, P20, P10) and tips.
11. Vortex mixer.
12. Microscope: With epifluorescence, 100 W mercury lamp, and appropriate filters (CHROMA Technology Corporation, Brattleboro, VT).
13. Charge-coupled devide (CCD) camera (Photometrics, Tucson, AZ).
14. Computer with software appropriate for FISH and CGH analysis (Applied Imaging Ltd., Santa Clara, CA; MetaSystems Group, Inc., Belmont, MA; Leica Microsystems Imaging Solutions Ltd., Bannockbun, Germany).
15. Formamide (Fisher Scientific, Fairlawn, NJ), cat. no. 227-500.
16. 100% Ethanol (200 proof).

17. Dextran sulfate (Pharmacia, Piscataway, NJ), cat. no. 17-0340-02.
18. Salmon sperm DNA: Sonicated (Pharmacia), cat. no. 27-4565-01.
19. RNase A (Sigma, St. Louis, MO), cat. no. 6513.
20. Cot-1™ DNA (Invitrogen, Carlsbad, CA), cat. no. 15279-011.
21. Proteinase K (Sigma), cat. no. P-6556.
22. IGEPAL (Sigma), cat. no. I-3021.
23. 20× Saline sodium citrate (SSC): 0.3 *M* NaCl, 0.3 *M* sodium citrate, pH 7.4. Combine 175.2 g of sodium chloride, 88.2 g of sodium citrate, and distilled water to a final volume of 1 L.
24. 2× SSC: Add 50 mL of 20× SSC to 450 mL of distilled water.
25. Denaturation solution: 70% formamide–2× SSC. Make fresh solution before each experiment. Combine 4 mL of 20× SSC, 8 mL of distilled water, and 28 mL of formamide. Adjust pH 7.0 with 1 *N* hydrochloric acid. Prewarm the denaturation solution to 70°C in a 70°C water bath.
26. RNase solution: 100 µg/mL in 2× SSC.
27. Phosphate-buffered solution (PBD; pH 8.0): Add 4 g of sodium bicarbonate and 20 mL of IGEPAL to 4 L of distilled water.
28. Hybridization buffer:
    a. For centromeric probes: 65% formamide, 2× SSC.
    b. For unique sequence probes: combine 5 mL of 50% formamide; 1 mL of 20× SSC, pH 7.0; 2 mL of 10% dextran sulfate; and 1 mL of salmon sperm DNA (10 mg/mL). Dilute to 10 mL with distilled water. Store at –20°C.
29. Ethanol: 70%, 85%, and 100% stored at room temperature and at –20°C.
30. 4,6,-diamidino-2-phenylindole (DAPI) counterstain (mutagen, avoid inhalation, ingestion, and contact with skin) (Vysis), cat. no. 32-804830.

3. Methods

3.1. FISH

An ISH protocol adheres to the following general outline:

1. Preparing slides.
2. Pretreating target (chromosomal) DNA on the slides.
3. Denaturating *in situ* target DNA.

4. Preparing probe.
5. ISH.
6. Post-hybridization washes.
7. Probe detection (immunocytochemistry).
8. Counterstaining.
9. Microscopy and image analysis.

3.1.1. Preparing Slides

Prepare metaphase chromosome spreads or interphase nuclei from fixed bone marrow or peripheral blood cell suspensions on glass microscope slides according to standard procedures. To achieve optimal results, use prepared slides within 1 wk. **Do not bake the slides.**

3.1.2. Pretreating Slides

3.1.2.1. RNase Treatment (Optional). RNase removes endogenous RNA and minimizes background signal.

1. Incubate the slides in DNase-free RNase solution (100 μg/mL in 2× SSC) for 1 h in a 37°C water bath.
2. Wash the slides in 2× SSC twice to remove excess RNase.
3. Dehydrate the slides in 70%, 85%, and 100% ethanol (2 min in each) at room temperature.
4. Air-dry the slides.

3.1.2.2. Proteinase K Treatment (Optional). Proteinase K increases the accessibility of the probe by digesting the chromosomal protein that surrounds the target nucleic acid.

1. Incubate the slides in 1 μg/mL of proteinase K solution for 3–5 min at 37°C. The time may vary for each slide.
2. Wash the slides twice in 2× SSC.
3. Dehydrate the slides in 70%, 85%, and 100% ethanol at room temperature for 2 min in each solution.
4. Air-dry the slides.

3.1.3. Denaturing In Situ Target DNA

Target chromosomal DNA can be denatured by alkaline (high pH) conditions or by heat.

1. Prewarm the denaturation solution (70% formamide–2× SSC) to 70°C in a water bath.
2. Denature slides for 2 min. *Time and temperature are important to maintain chromosome morphology. For every slide, there will be a decrease of 1°C.*
3. Immediately transfer slides to a Coplin jar containing 40 mL of ice-cold 70% ethanol. Rinse slides for 2 min. Repeat rinses for 2 min each in cold 85% ethanol, then cold 100% ethanol.
4. Allow slides to air-dry or dry under an air jet.

3.1.4. Preparation of Probe

This subheading gives general guidelines; the manufacturer's recommendations should be followed whenever applicable.

3.1.4.1. Specific Centromeric Probes (Alpha, Beta, or Classical).

1. Prewarm the tube containing the probe at 37°C for 5 min, then vortex-mix and centrifuge 2–3 s to collect the contents in the bottom of the tube.
2. Combine 1.5 µL of biotin- or digoxigenin-labeled probe with 30 µL of hybridization buffer (65% formamide, 2× SSC) in a 0.5-mL microcentrifuge tube.
3. Denature the probe in 70°C water bath for 5 min. Chill quickly on ice. Centrifuge 2–3 s.

3.1.4.2. Whole Chromosome Painting Probes/Arm Painting Probes. Because the painting probes contain repeat sequences, Cot-1 DNA is added to the probe mixture, and the mixture is incubated before hybridization. The Cot-1 DNA binds to the repeat sequences in the probe, leaving the unique probe sequences to hybridize with the target DNA. *Some of the direct-labeled probes do not require preannealing.*

1. Prewarm probe at 37°C for 5 min, vortex-mix, and centrifuge 2–3 s to collect the contents in the bottom of the tube.
2. Remove 15 µL of probe to a microcentrifuge tube.
3. Denature the probe at 70°C for 10 min, then centrifuge 2–3 s.
4. Incubate in a 37°C water bath for 1–2 h to preanneal the repetitive sequences. Centrifuge for 2–3 s.

3.1.4.3. Unique Sequence/Single Copy Probes. Follow the manufacturer's instructions when applicable. For home brew probes follow the steps given below.

1. Mix 100–200 ng of the labeled probe with 28 µL of hybridization buffer and 1 µL of Cot-1 DNA.
2. Denature the probe at 70°C for 7 min.
3. Chill the probe on ice immediately after denaturation.

3.1.4.4. Subtelomeric Probes.

1. Add 1 µL of the probe, 7 µL of the hybridization buffer, and 1 µL of water into a microcentrifuge tube. Vortex-mix and centrifuge for few seconds.
2. Denature at 72°C for 5 min, and then centrifuge for 2–3 s.

3.1.5. In Situ *Hybridization*

1. Place the denatured probe mix on the slide with the denatured chromosomal (target) DNA and cover with a glass coverslip.
2. Seal by applying rubber cement along the perimeter of the coverslip to prevent evaporation of the hybridization buffer.
3. Incubate at 37°C in a humidified chamber for 4–16 h.

3.1.6. Post-Hybridization Washes

Unhybridized and nonspecifically bound probe is removed by washes of various stringencies. The stringency of these washes can be modified by varying the temperature, as well as the concentrations of formamide and salt. Higher stringency is achieved by increasing the temperature, decreasing the salt solution, or increasing the formamide concentration. For example, repeat sequence probes are washed at a higher stringency, whereas all other types of probes are washed at a lower stringency. *For direct-labeled probes, reduce the time of washes in formamide and SSC by half or more.*

1. Prewarm the 50% formamide–2× SSC wash solution to 43–45°C for 30 min.
2. Remove the rubber cement and place the slides in a Coplin jar containing the prewarmed 50% formamide–2× SSC. Incubate for 10 min. The coverslips will fall off.
3. Wash the slides two times in 50% formamide at 43°C for 5 min each time.
4. Wash the slides four times in 2× SCC at 43°C for 5 min each time.
5. Place the slides in PBD to rinse off excess salt and formamide until the probe detection.
6. Slides can be stored at 4°C for up to 2 wk.

3.1.7. Probe Detection

Remove the slides from PBD and blot excess fluid. *Do not allow the slides to dry.* Follow the manufacturer's instruction when applicable.

3.1.7.1. DETECTION OF BIOTIN-LABELED PROBES. Biotin-labeled probes can be visualized by using American Laboratory Technologies Inc. (cat. no. HK100-2), Cytocell (cat. no. ADR 001, ADR 002, and ADR 003), Cambio FITC (cat. no. CA-1066K), or Texas Red–biotin (cat. no. CA-C-1082-KT) detection kits. When using Cytocell detection kits, follow the steps outlined under digoxigenin probes. The biotin-labeled probes will be detected as red (Cy3) by the Cytocell detection kit.

1. Apply 50 µL of FITC–avidin, cover with a plastic coverslip, and incubate 30 min at 37°C. Remove coverslip, then wash slides three times (2 min each time) in 1× PBD at room temperature.
2. Apply 20 µL of DAPI (0.5 µg/mL in antifade solution) to the slide and cover with a glass coverslip. View under a fluorescence microscope.

If the signal is weak, perform the following steps to amplify the signal.

3. Remove the coverslip and perform three 2-min washes in PBD.
4. Apply 50 µL of anti-avidin–FITC antibody and incubate for 15 min at 37°C. Repeat the washes in PBD.

5. Apply 50 µL of FITC–avidin conjugate and incubate for 15 min at 37°C. Repeat the washes in PBD.

3.1.7.2. DETECTION OF DIGOXIGENIN-LABELED PROBES. *Digoxigenin-labeled probes can be detected by using the Cytocell dual-color detection reagents. These kits detect digoxigenin-labeled probes by their green color (FITC).*

1. Apply 50 µL of mouse anti-digoxin–FITC conjugate mixed with streptavidin conjugated to Cy3 (Cytocell, cat. no. ADR 001) to the slide, cover with a coverslip, and incubate at 37°C for 15 minutes.
2. Remove the coverslip and perform three 2-min washes in PBD.
3. Add 20 µL of DAPI and view under a fluorescence microscope.

If the signal is weak, perform the following steps to amplify the signal.

4. Add 50 µL of rabbit anti-mouse–FITC conjugate mixed with biotinlyated antistreptavidin anti-sheep antibody (Cytocell, cat. no. ADR 002) and incubate for 15 min at 37°C. Repeat the washes in PBD.
5. Apply 50 µL of goat anti-rabbit–FITC conjugate mixed with streptavidin conjugated to Cy3 (Cytocell, cat. no. ADR 003) and incubate for 15 min at 37°C. Repeat the washes in PBD.

3.1.8. Counterstaining

The cells are recognized by using counterstains such as propidium iodide and DAPI. The use of antifading agents such as diphenylene diamine will preserve the signals during storage and image acquisition.

1. Add 20 µL of the counterstain to the slide and cover with a glass coverslip.

3.1.9. Microscopy and Image Analysis

Using appropriate filters, take photographs with Kodacolor 400 and Fujichrome 400 film. Digital imaging systems are now widely used for FISH analysis. The imaging system consists of a combination of a digital or CCD camera and a computer with advanced software (Applied Imaging, Metasystems, etc.).

3.2. Comparative Genomic Hybridization

3.2.1. Reagents

1. Nick translation kit (Invitrogen, Carlsbad, CA), cat. no. 18160-010.
2. Sephadex G-50 DNA grade F nick spin columns (Pharmacia), cat. no. 17-0855-02.

3.2.2. Metaphase Spreads

Prepare metaphase spreads of phytohemagglutinin (PHA)-stimulated lymphocytes from healthy subjects by using standard cytogenetic procedures with hypotonic treatment and methanol–acetic acid fixation.

3.2.3. Labeling of Tumor and Normal DNA

3.2.3.1. TUMOR DNA. High-molecular-weight tumor DNA for analysis is labeled by nick translation as follows:

1. In a microcentrifuge tube, combine 1 µg of tumor DNA, 5 µL of 10× A4 mixture (0.2 m*M* each of dATP, dCTP, and dGTP and 0.1 m*M* dTTP in 500 m*M* Tris-HCI, pH 7.8; 50 m*M* $MgC1_2$; 100 m*M* β-mercaptoethanol; and 100 µL/mL of bovine serum albumin), 1 µL of biotin–14-dUTP, 5 µL of enzyme mixture (containing 0.5 U/µL of DNA polymerase I and 0.04 U/µL of DNase I), and 1 µL (10 U/µL) DNA polymerase I (Invitrogen, cat. no. 18010-017). Dilute to 50 µL with distilled water.
2. Incubate for 45–60 min at 15°C in a water bath.
3. Stop the reaction by incubating for 10 min at 70°C.

3.2.3.2. NORMAL DNA. The reference (normal) DNA is labeled as described in **Subheading 3.2.3.1.** except that digoxigenin-11-dUTP is used instead of biotin-14-dUTP. Alternatively, the reference and tumor DNA can be labeled with direct fluorochromes (e.g., Spectrum Orange and Spectrum Green).

3.2.3.3. DETERMINING THE SIZE OF DNA. Determine the size of the tumor and normal DNA by electrophoresis through a 1% agarose gel. The size of the DNA should range from about 500 to 2000 basepairs.

The fragment length can be modified by adjusting the ratio of DNase to DNA polymerase in the nick translation reaction or by varying the incubation time. The labeled DNA samples are separated from unincorporated nucleotides on Sephadex G-50 column.

3.2.3.4. Precipitation and Denaturation of Tumor and Normal DNA.

1. In a microcentrifuge tube, mix 200 ng each of the labeled tumor and normal DNA, 10–15 µg of unlabeled Cot-1 DNA (to block the binding of repetitive sequences), and 3 µL of 3 *M* sodium acetate. Dilute to 100 mL with ethanol and store at –20°C for 2–16 h.
2. Precipitate the DNA by centrifugation at 11,000*g* for 30 min at 4°C.
3. Remove the supernatant and air or vacuum dry the pellet.
4. Dissolve the dried pellet in 10 µL of hybridization buffer.
5. Denature the probe in a water bath at 70°C for 5 min.

3.2.4. Denaturation and Preparation of Slides

1. Denature metaphase chromosome spreads at 70°C in 70% formamide–2× SSC for 2.5 min.
2. Dehydrate slides in 70%, 85%, and 100% cold ethanol solutions (2 min in each). Air-dry.
3. Incubate the slides in 0.1 µg/mL of proteinase K solution for 5.5 min at room temperature.
4. Wash three times (2 min each time) in 2× SSC.
5. Dehydrate in 70%, 85%, and 100% ethanol. Air-dry.

3.2.5. Hybridization

1. Add the denatured probe to the denatured slide and hybridize for 2 d at 37°C in a humidified chamber.

3.2.6. Post-Hybridization Washes

For indirect labeled probes follow the steps below.

1. Remove the coverslip. To remove unbound DNA, wash the slides three times (5 min each time) in 50% formamide–2× SSC, pH 7.0, at 45°C.
2. Wash twice in 2× SSC and once in 0.1× SSC at 45°C (5 min each wash).

3. Wash the slides three times in PBD for 2 min.
4. Apply 50 µL of rhodamine antidigoxigenin–FITC avidin, or Cytocell dual color detection reagents, cover with a coverslip, and incubate at 37°C for 15 min. Remove the coverslip and wash slides three times in PBD for 2 min.
5. Counterstain with DAPI.

3.2.7. Digital Image Acquisition and Processing

1. Acquire blue, red, and green images by using a quantitative image processing system with a fluorescence microscope that is equipped with a cooled CCD camera and appropriate filter sets. The software program integrates the green and red fluorescence intensity in stripes orthogonal to the chromosomal axis, subtracts local background, generates the intensity profiles for red and green colors, and calculates the ratio profiles for both colors from the p-terminus to the q-terminus of each chromosome.
2. Acquire 5–10 metaphases from each sample. Average the ratio profiles from each chromosome type. The ratio profiles for all chromosomes can be used to generate the "copy number karyotype" of the tumor cells.
3. To obtain the normal thresholds for the analysis of tumor/normal hybridization, perform control hybridization with normal/normal DNA. A ratio of tumor DNA intensity to control DNA intensity > 1.25 is generally considered to indicate a gain of chromosomal regions, and a ratio < 0.85 is considered to indicate a loss.

3.3. Spectral Karyotyping

3.3.1. Equipment Required

1. Epifluorescence microscope attached with a C-mount interface to a dual-mode optical head one with a Fourier transform spectrometer (Sagnac common path interferometer) and a high-performance, 12-bit digital CCD camera (Princeton Instruments, Trenton, NJ).
2. Xenon lamp (OptiQuip 770/1600)
3. Custom-designed filter set (Chroma Technology, Brattleboro, VT).
4. SD-200 Spectracube™ spectral bio-imaging system (ASI).

3.3.2. Reagents Required

1. SKY ASR1001H WCP probes.
2. Blocking reagent (vial 2 from ASI).
3. Cy5 staining reagent (vial 3 from ASI).
4. Cy5.5 staining reagent (vial 4 from ASI).
5. Pepsin, 10% stock solution (Sigma), cat. no. P6887.
6. 1 *M* $MgCl_2$ (Sigma), cat. no. M1028.
7. 37% Formaldehyde (Sigma), cat. no. F1268.
8. Anti-fade DAPI reagent (vial 5 from ASI).
9. 12 *N* HCl.
10. Phosphate-buffered saline (PBS) (Sigma), cat. no. 1000-3.

3.3.3. Preparation of Reagents

1. Washing solution I (50% formamide–2× SSC). Mix 20 mL of formamide, 4 mL of 20× SSC, and 16 mL of distilled H_2O in a Coplin jar.
2. Washing solution II (1× SSC). Add 25 mL of 2× SSC to 25 mL of distilled water. Mix well and heat to 45°C.
3. Washing solution III (PBD). (*See* FISH procedure, **Subheading 2.**)
4. 0.01 *M* HCl. Add 83 µL of 12N HCl to 50 mL of distilled water. Heat to 37°C in a glass Coplin jar.
5. Pepsin stock solution. Prepare a 10% pepsin stock solution (100 mg/mL) in sterile water. Dissolve completely and aliquot. Store at –20°C.
6. Proteinase K stock solution. Prepare a 5 mg/mL solution of proteinase K. Dissolve completely and aliquot. Store at –20°C.
7. 1× PBS–$MgCl_2$. Add 50 mL of 1 *M* $MgCl_2$ to 950 mL of 1× PBS.
8. 1% Formaldehyde. Add 2.7 mL of 37% formaldehyde to 100 mL of 1× PBS–$MgCl_2$.
9. Denaturing solution: 70% formamide in 2× SSC.

3.3.4. Day 1 (Check list before starting experiment.)

1. Turn on the water baths at 37°C and at 72°C.
2. Prewarm the denaturing solution to 72°C.
3. Prewarm 2× SSC in a water bath to 37°C.
4. Prewarm the humidified chamber in the incubator.
5. Alcohol at room temperature and at –20°C.

3.3.5. Selection and Pretreatment of Slides

Select the best slide without cytoplasm for hybridization. The slides should be aged at room temperature for 3–5 d. Slides can be pretreated with either pepsin or proteinase K. Pretreatment to remove residual cytoplasm is an important step, but overtreatment with pepsin or proteinase K can lead to overdigestion and poor chromosome morphology. The pretreatment time should be adjusted to remove the cytoplasm completely.

3.3.5.1. Slide Pretreatment with Pepsin.

1. Prewarm 50 mL of 0.01 *M* HCl to 37°C in a glass Coplin jar. Add 5–15 µL of pepsin stock solution and mix well. Incubate slides at 37°C in the pepsin solution for 3–5 min.
2. Wash slides in 1× PBS at room temperature for 5 min.
3. Repeat 1× PBS wash for 5 min.

3.3.5.2. Slide Pretreatment with Proteinase K.

1. Add 5 µL of proteinase K stock solution to 50 mL of distilled water and warm to 37°C in a water bath. Incubate the slides in the solution at 37°C for 4–7 min. The time must be optimized for each slide.
2. Wash slides in 1× PBS at room temperature for 5 min.
3. Repeat 1× PBS wash for 5 min.

Note: *Do not dry the slides during this step. Use PBS and coverslip and check under a phase-contrast microscope for any remaining cytoplasm. If cytoplasm remains, repeat treatment.*

4. Wash slides in 1× PBS–$MgCl_2$ at room temperature for 5 min.
5. Place slides in a Coplin jar containing 1% formaldehyde and incubate for 10 min at room temperature.
6. Wash in 1× PBS for 5 min.
7. Dehydrate in 70%, 85%, and 100% ethanol (2 min in each) at room temperature. Air-dry the slides.

3.3.6. Denaturation of Chromosomes

1. Place slides in 70% formamide–2× SSC, pH 7.0, at 72°C for 1.5 min. **Do not overdenature.**

2. Immediately dehydrate in 70%, 85%, and 100% **ice-cold** ethanol (2 min in each).
3. Air-dry the slides.

3.3.7. Probe Denaturation

1. Prewarm probe with the hybridization mixture at 37°C for 5 min. Vortex gently and centrifuge 2–3 s. Place 10 µL of the probe (for half slide) in a microcentrifuge tube.
2. Denature the probe in a water bath at 80°C for 7 min, and then centrifuge 2–3 s.
3. Preanneal the probe in a water bath at 37°C for 60 min.
4. Centrifuge for a few seconds.

3.3.8. Hybridization

1. Apply the denatured probe to the denatured chromosomes on the slide.
2. Apply glass coverslip and seal the edges with rubber cement; incubate at 37°C in a humidified chamber for two nights.

3.3.9. Post-Hybridization Wash (D 3)

During the entire procedure, the slide should remain wet and protected from direct light.

Formamide wash:

1. Remove the rubber cement and coverslip. Place slides in 50% formamide–2× SSC, pH 7.0, at 43–45°C for 5 min. Agitate the slide occasionally.
2. Apply 50 µL of blocking reagent (vial 2), place a plastic coverslip and incubate at 37°C for 30 min.
3. Wash the slide in washing solution II (1× SSC, pH 7.0) at 43–45°C for 5 min.
4. Transfer to washing solution III (PBD) at room temperature and proceed with detection.

3.3.10. Detection

1. Remove the coverslip and allow the fluid to drain. Apply 50 µL of Cy5 staining reagent to the slide, cover with a plastic coverslip, and incubate for 45 min at 37°C in a humidified chamber.

2. Rinse three times (2 min each time) in washing solution III (PBD) at room temperature.
3. Apply 50 µL of Cy5.5 staining reagent, cover with a plastic coverslip, and incubate at 37°C for 45 min.
4. Rinse three times (2 min each time) in washing solution III (PBD) at room temperature.
5. Wash slides briefly in water and air-dry.
6. Add 20 µL of the anti-fade–DAPI reagent (vial 5, ASR 1005H) and cover with a glass coverslip.

3.3.11. Image Acquisition and Analysis

For best results, acquire images within a week after the detection of probes. Spectral and DAPI images are captured separately for each metaphase. Both the spectral and DAPI images are analyzed by using the SkyView™ software. The principles of spectral imaging and analysis are explained in Schrock et al. ***(3)***.

4. Advantages and Disadvantages of Fish, CGH, and Sky Techniques

FISH is a very rapid, sensitive, and cost-effective technique that offers the capability to detect both numerical and structural chromosomal abnormalities in interphase and metaphase nuclei. FISH also allows simultaneous detection of multiple, differentially labeled probes and the quantification of hybridization signals. FISH permits rapid sex determination, detection of minimal residual disease, follow-up of patients who receive sex-mismatched bone marrow transplants, and detection of early relapse ***(16)***. For example, the cryptic t(12;21)(p13;q22), the most common genetic abnormality observed in childhood B-lineage ALL, is not detected by conventional cytogenetics. However, FISH can identify this translocation both in interphase and metaphase nuclei using probes for the *TEL-AML1 (ETV6-CBFA2)* genes. FISH can also identify new partner breakpoints of known genes, and the partner gene(s) involved in a translocation can then be cloned ***(17,18)***. However, FISH has a number of limitations, including cross-hybridization of nonspecific fluorescent signals, nonspecific background, and suboptimal signal intensity. FISH with painting probes cannot detect small interstitial deletions, duplications, or inversions. Although the sensitivity of interphase FISH for specific fusion transcripts is less than that of RT-PCR, it can detect deletions and numerical abnormalities that are not detected by RT-PCR.

CGH allows the entire genome to be viewed at a glance and permits a relatively rapid and accurate assessment of genetic abnormalities in tumor cells. Unlike conventional cytogenetics and FISH, CGH requires only the genomic DNA, thus allowing the use of archived specimens. Also, unlike FISH, CGH does not require prior knowledge of the genomic region to be studied. CGH can detect copy number changes and gains and losses of chromosomal regions. CGH has been used to evaluate ALL *(19–20)*; however, its utility is limited mostly to cases that have hyperdiploid chromosomes. CGH has several limitations. It does not detect chromosomal balanced translocations, inversions, and intragenic rearrangements *(21)*. Also, CGH does not provide information about the identity of the amplified or deleted segments or the arrangement of these regions in marker chromosomes of the test genome. One important limitation of CGH is that differences in copy number can be detected only if the sample contains more than 50% abnormal cells. Furthermore, CGH is not appropriate for studying the clonal heterogeneity of a neoplastic sample, nor can it reveal the ploidy of different cells in the sample. The sensitivity of the technique in detecting low copy number increases or decreases is in the range of 10–20 Mb *(22)*. The detection limit of amplification is 2 Mb *(23)*.

SKY can successfully refine complex karyotypes and detect hidden structural abnormalities, thus revealing new recurrent translocations *(24,25)*. As multicolor karyotyping identifies novel translocations that better characterize the disease entities, the information obtained may in turn improve the diagnosis, treatment stratification, and prognosis of these diseases. In spite of the great advantages, SKY can be performed only on metaphase spreads. SKY does not detect small rearrangements or intrachromosomal rearrangements such as deletions, duplications, or inversions. Previous studies have shown that the sensitivity of the SKY technique is in the range of 1.5 Mb *(3,26)*. However, the consistent failure to detect t(12;21) in ALL shows the sensitivity of SKY to be < 1.5 Mb. Also, SKY analysis requires very sophisticated equipment and expensive probe.

In summary, the combination of conventional cytogenetics with different molecular cytogenetic techniques improves the accuracy of detecting chromosomal abnormalities and provides valuable information on the risk stratification of pediatric ALL.

References

1. Kallioniemi, A., Kallioniemi, O-P., Sudar, D., et al. (1992) Comparative genomic hybridization for molecular cytogenetic analysis of solid tumors. *Science* **258,** 818–821.
2. Speicher, M. R., Ballard, S. G., and Ward, D. C. (1996) Karyotyping human chromosomes by combinatorial multifluor FISH. *Nat. Genet.* **12,** 368–375.
3. Schrock, E., du Manoir, S., Veldman, T., et al. (1996) Multicolor spectral karyotyping of human chromosomes. *Science* **273,** 494–497.
4. Tanke, H. J., Wiegant, J., van Gijlswijk, R. P., et al. (1999) New strategy for multicolor fluorescence in situ hybridization: COBRA: COmbined Binary RAtio Labeling. *Eur. J. Hum. Genet.* **7,** 2–11.
5. Henegariu, O., Heerema, N. A., Bray-Ward, P., and Ward, D. C. (1999) Color changing karyotyping: an alternative to M-FISH/SKY. *Nat. Genet.* **23,** 263–264.
6. Muller, S., O'Brien, P. C., Ferguson-Smith, M. A., and Wienberg, J. (1998) Cross-species color segmenting: a novel tool in human karyotype analysis. *Cytometry* **33,** 445–452.
7. Chudoba, I., Plesch, A., Lorch, T., Lemke, J., Claussen, U., and Senger, G. (1999) High resolution multi-color banding: a new technique for refined FISH analysis of human chromosomes. *Cytogenet. Cell Genet.* **84,** 156–160.
8. Rigby, P. W., Dieckmann, M., Rhodes, C., and Berg, P. (1997) Labeling of deoxyribonucleic acid to high specific activity in vitro by nick translation with DNA polymerase I. *J. Mol. Biol.* **113,** 237–251.
9. Collins, C., Kuo, W. L., Segraves, R., Fuscoe, J., Pinkel. D., and Gray, J. W. (1991) Construction and characterization of plasmid libraries enriched in sequences from single human chromosomes. *Genomics* **11,** 997–1006.

10. Vooijs, M., Yu, L. C., Tkachuk, D., Pinkel, D., Johnson, D., and Gray, J. W. (1993) Libraries for each human chromosome, constructed from sorter-enriched chromosomes by using linker-adaptor PCR. *Am. J. Hum. Genet.* **52,** 586–597.
11. Guan, X. Y., Zhang, H., Bittner, M., Jiang, Y., Meltzer, P., and Trent, J. (1996) Chromosome arm painting probes. *Nat. Genet.* **12,** 10–11.
12. Ning , Y., Roschke, A., Smith, A. C.M., et al. (group 1), Flint, J., Horsley, S., Regan, R., et al. (group 2) (1996) A complete set of human telomeric probes and their clinical application. *Nat. Genet.* **14,** 86–89.
13. Ballif, B. C., Kashork, C. D., and Shaffer, L. G. (2000) FISHing for mechanisms of cytogenetically defined terminal deletions using chromosome-specific subtelomeric probes. *Eur. J. Hum. Genet.* **8,** 764–770.
14. Knight, S. J. and Flint, J. (2000) Perfect endings: a review of subtelomeric probes and their use in clinical diagnosis. *J. Med. Genet.* **37,** 401–409.
15. Telenius, H., Carter, N. P., Bebb, C. E., Nordenskjold, M., Ponder, B. A., and Tunnacliffe, A. (1992) Degenerate oligonucleotide-primed PCR: general amplification of target DNA by a single degenerate primer. *Genomics* **13,** 718–725.
16. Le Beau, M. (1993) Fluorescence in situ hybridization in cancer diagnosis, in *Important Advances in Oncology* (DeVita, V. T., Hellman, S., and Rosenberd, S. A., eds.), J. B. Lippincott, Philadelphia, pp. 29–45.
17. Romana, S. P., Le Coniat, M., and Berger, R. (1994) t(12;21): a new recurrent translocation in acute lymphoblastic leukemia. *Genes Chromosomes Cancer* **9,** 186–191.
18. Romana, S. P., Poirel, H., Leconiat, M., et al. (1995) High frequency of t(12;21) in childhood B-lineage acute lymphoblastic leukemia. *Blood* **86,** 4263–4269.
19. Larramendy, M. L., Huhta, T., Heinonen, K., et al. (1998) DNA copy number changes in childhood acute lymphoblastic leukemia. *Hematologica* **83,** 890–895.
20. Haas, O., Henn, T., Romanakis, K., du Manoir, S., and Lengauer, C. (1998) Comparative genomic hybridization as part of a new diagnostic strategy in childhood hyperdiploid acute lymphoblastic leukemia. *Leukemia* **12,** 474–481.
21. Green, G. A., Schrock, E., Veldman, T., et al. (2000) Evolving molecular cytogenetic technologies, in *Medical Cytogenetics* (Mark, H. F.L., ed.), Marcel Dekker, New York, pp. 579–592.

22. Benz, M., Plesch A., Stilgenbauer, S., Dohner, H., and Lichter, P. (1998) Minimal sizes of deletions detected by comparative genomic hybridization. *Genes Chromosomes Cancer* **21,** 172–175.
23. Piper, J., Rutovitz, D., Sudar, D., et al. (1995) Computer image analysis of comparative genomic hybridization. *Cytometry* **19,** 10–26.
24. Veldman, T., Vignon, C., Schrock, E., Rowley, J. D., and Ried, T. (1997) Hidden chromosome abnormalities in hematological malignancies detected by multicolor spectral karyotyping. *Nat. Genet.* **15,** 406–410.
25. Mathew, S., Rao, P. H., Dalton, J., Downing, J. R., and Raimondi, S. C. (2001) Multicolor spectral karyotyping identifies novel translocations in childhood acute lymphoblastic leukemia. *Leukemia* **3,** 468–472.
26. Haddad, B. R., Schrock, E., Meck, J., et al. (1998) Identification of de novo chromosomal markers and derivatives by spectral karyotyping. *Hum. Genet.* **103,** 619–625.

16

Solving Problems in Multiplex FISH

Jon C. Strefford

1. Introduction

Over the last 15 yr, advances in molecular biology have allowed improvements in the sensitivity and versatility of cytogenetic analysis. These advances have included developments in recombinant technology such as fluorescence *in situ* hybridization (FISH), a means of detecting chromosome rearrangements through the use of DNA-specific probes known as chromosome paints. Recent extensions to this painting technology are multiplex FISH (MFISH) *(1)* and spectral karyotyping (SKY) *(2)*. These technologies use complex combinatorial probes and sophisticated image analysis software to allow each of the 24 different human chromosomes to be simultaneously identified in 24 discrete colors, making it possible to screen for rearrangements between all chromosomes in one analysis. These techniques have been used to characterize hematological malignancies *(3)* and bladder *(4,5)* and prostate cancer *(6–8)*. MFISH represents a useful tool for the genome-wide screening of chromosome rearrangement and is becoming increasingly practiced in both the research and diagnostic professions. It follows that a chapter on the use of these technologies is appropriately included in a series on the methods used in malignancy cytogenetics.

From: *Methods in Molecular Biology, vol. 220: Cancer Cytogenetics: Methods and Protocols*
Edited by: John Swansbury © Humana Press Inc., Totowa, NJ

Although the MFISH methodology is similar in principle to standard FISH protocols, the dynamics of the probe binding and image analysis are far more complex. Therefore, technical variations and deviations that are insignificant in conventional FISH may be critical to the intricacies of MFISH. This chapter highlights several key steps in the entire process of MFISH, beginning with the setting up of the sample and culminating with the computer analysis. Several technical changes that may optimize both the quality of the MFISH results and their reliability are then discussed.

2. Materials

Only solutions and materials directly required for the MFISH procedure are listed. The materials used for the processing of samples to obtain metaphase cells are as described in other chapters in this book.

Note: *Chemicals indicated by (*) are known or potential carcinogens/poisons and should be handled with due care and attention.*

1. Containers and pipets: 0.5-mL Eppendorf tubes, Falcon tubes, plastic disposable and glass pipets.
2. Fixative: Three parts absolute methanol and one part glacial acetic acid. This should be freshly prepared and chilled to –20°C.
3. Ethanol: 70%, 95%, and 100% ethanol series. Can be made up and stored in sealed containers at room temperature for up to 3 mo.
4. 20× Saline sodium citrate (SSC): 175.3 g of NaCl and 88.2 g of sodium citrate made up to 1 L, adjusted to pH 7.0–7.5 with 1 *M* HCl or 1 *M* NaOH. Solution can be stored at room temperature for approx 2 m or at 4°C for up to a year.
5. FISH probe: In this laboratory, SpectraVysion MFISH probe (Vysis, UK) is used. This probe can be stored in the dark at –20°C for approx 18 mo.
6. FISH denaturation solution: 70% Formamide (*)–2× SSC, pH 7.0–7.5: 35 mL of formamide, 5 mL of 20× SSC, and 10 mL of distilled water. Check pH again; adjust to 7.0–7.5.
7. FISH post-hybridization wash solutions: (a) 0.4× SSC + 0.3% Nonidet P-40 (NP-40) (Sigma, cat. no. I-3021), at pH 7.0–7.5: 5 mL 20× SSC, 0.15 mL of NP-40, 43 mL of distilled water. Check pH, then make up to 50 mL with distilled water. (b) 2× SSC+0.1% NP-

40, pH 7.0– 7.5: 5 mL of 20× SSC and 43 mL of distilled water + 0.05 mL of NP-40. Check pH, adjust to 7.0–7.5, and make up to 50 mL with distilled water.

8. Stain and anti-fade solution: 4', 6-diamidino-2-phenylindole (DAPI) counterstain (*) and Citifluor anti-fade. In this laboratory a relatively dilute solution is used (1 µL of DAPI + 4999 µL of Citifluor anti-fade). This solution can be stored in the dark at 4°C until required.
9. Coverslips: 22 × 22 mm, grade 1.5 are acceptable. In this laboratory 19-mm circular coverslips are used to reduce the amount of probe that is required.
10. Slides: Frosted-end slides are preferable for convenience of labeling. The slides must be free of grease and dirt. In this laboratory slides are washed in 1% lypsol detergent for approx 15 min followed by a 30-min rinse in running tap water. The slides are then stored in distilled water at 4°C until required. These slides can be kept in the refrigerator for 1–2 h before use.
11. Rubber cement: Rubber cement is used to seal the probe under the coverslip, so that the hybridization step can be carried out. In this laboratory, Cow gum is used, although virtually any sealant is acceptable.

3. Methods

3.1. The MFISH Probe

The MFISH probe is photoactive and therefore should be stored in the dark at –20°C. All the practical steps using this type of probe should be done in subdued light. When the probe is required, briefly vortex-mix and centrifuge before aliquoting the required amount. In this laboratory half the recommended volume of probe is used. This does not affect the result. It can be used pure or diluted with an equal volume of hybridization buffer.

MFISH probes are generally produced using five fluorochromes so that each chromosome paint is labeled with a unique combination. Computer software can then combine these images and assign a pseudo-color depending on the combination. In this laboratory the SpectraVysion MFISH probe is used (Vysis, UK). This probe utilizes the Spectrum Gold, Spectrum Far-Red, Spectrum Red, Spectrum Green, and Spectrum Aqua fluorochromes.

3.2. Choice of Cultures for MFISH

The cultures available for MFISH analysis depend on the type of sample received and the desired cell type needed for analysis. The main application of MFISH is to permit further characterization of chromosome abnormalities in hematological disorders and cancer, as chromosome anomalies detected in pre- and post-natal samples are likely to be relatively simple and not justify the use of such an expensive technique. Therefore, standard methodologies used for the culturing of cells from malignant disease are of particular importance. The most important factor in selecting cultures for MFISH is a relatively high mitotic index, as MFISH probe is expensive and several metaphases must be analyzed to assess the possibility of karyotypic evolution or even the presence of separate malignant clones. Also worth considering is the length of the metaphase chromosomes, as complex karyotypes involving small chromosome rearrangements may be difficult to define. A synchronized (blocked) culturing procedure may therefore be desirable. However, good spreading and chromosome separation is also very important, as touching or overlapping chromosomes will produce flaring, spurious new colors caused by new combination ratios of the fluorochromes, which interferes with the analysis (*see* **Note 1**).

3.3. Slide Preparation

Slides for MFISH analysis are prepared in a similar way to those for standard cytogenetic analysis, although the following adaptations may improve the quality of the slides:

1. The concentration of the cell suspension is of great importance, and judging the correct dilution will come with experience. If concentrations are too high the chromosomes will not spread properly, whereas if the suspension is too dilute, inadequate cells for analysis will be obtained and valuable probe may be wasted. In this laboratory slides are cleaned in 1% lypsol solution, stored in water, and used when wet. However, slides cleaned with fixative and then air-dried have also been used successfully.

2. One or two drops of cell suspension onto a slide will be adequate. Spreading may be aided by the addition of a further drop of fresh fix. Assessing the quality of each slide using phase-contrast microscopy is essential to ensure that there will be suitable metaphases available before committing time and expensive reagents to it.
3. The two most important factors in avoiding problems with MFISH are the quantity of cytoplasm and the position of target metaphases in relation to nearby interphase nuclei. Levels of cytoplasm around a target metaphase need to be as low as possible because high levels may influence probe access. In extreme cases, high-cytoplasm preparations may completely disrupt MFISH analysis by inhibiting the visualization of several of the fluorochromes, in particular the red and far-red photoactive dyes. Metaphases that are not proximal to any interphase cells seem to be the best target for MFISH, as chromosomes near to interphase cells tend to respond poorly to the MFISH procedure. This may influence the MFISH result either by affecting levels of cytoplasm and hence probe binding, or by affecting image capture through the intensity of a nearby painted interphase cell. If the chromosome preparations contain high levels of cellular debris, a water fixation step may be useful (one part water + three parts fixative, centrifuge), followed by some changes of fresh fixative before spreading a fresh slide.

3.4. MFISH Procedure

Each MFISH probe kit contains a detailed manufacturer's protocol, which must be followed precisely. These tend to assume that everything will work perfectly the first time; because this is not always the case, some helpful pointers are described in the following.

1. Slide ageing: After slide preparation, drying on a hotplate at approx. 65°C for 2 h, followed by leaving overnight at room temperature, helps to age the chromosomes. It is important to age the slide as this preserves the chromosome morphology and aids the hybridization and inverted DAPI banding. It is useful at this stage to mark the required region of the slide with a diamond-tipped pen.
2. Slide pretreatment: In our experience, pretreatment steps are not necessary, although several standard FISH pretreatment protocols are commonly used in other laboratories prior to the MFISH reaction. However, an incubation in 2× SSC (45 min at 37°C) can help remove some cellular debris and improve the inverted DAPI banding.

3. Making up FISH solutions: pH is critical to the good visualization of the red fluorochromes, in particular the far-red. The pH should be checked after each solution has been made up.
4. After denaturation and subsequent dehydration it can be advantageous to place the slide on a hotplate (46°C) to stop the slide reaching hybridization temperature before the probe and coverslip are added and sealed.
5. A co-denaturing step can also be employed for the MFISH procedure; it results in better chromosome morphology and improves the inverted DAPI banding. This can be achieved by placing the probe directly onto the slide before both are simultaneously denatured (5 min at 74°C).

3.5. Image Capture

Specialized MFISH software, as well as a specific filter wheel system, are required for MFISH capture and analysis. The technical information can be obtained from the various specialized cytogenetic imaging companies that supply these computer systems. An example of an MFISH karyotype is shown in **Fig. 1**.

1. When looking for MFISH metaphases of high quality for capture, it is critical that the scanning is not performed using the DAPI filter. This is because 'bleed' through the DAPI filter will fade many of the other fluorochromes. In this laboratory scanning is performed through the Spectrum Gold filter at high power (×100), as this gives a good image without disrupting other channels.
2. Conventional FISH images can be captured with standard xenon light sources. MFISH images can also be captured with this source of fluorescence, but considerably superior results can be obtained using a combination xenon/mercury system. The disadvantage of this xenon/mercury light source is the intensity of the fluorescence, which often burns out several of the fluorochromes after only a single image capture. The red and far-red are particularly sensitive to this light source. Therefore, it is important to capture each image on the first attempt, as it will generally not be possible to capture the red and far-red fluorochromes a second time. Achieving this will come with experience, although the auto-capture option on most MFISH systems should be adequate.

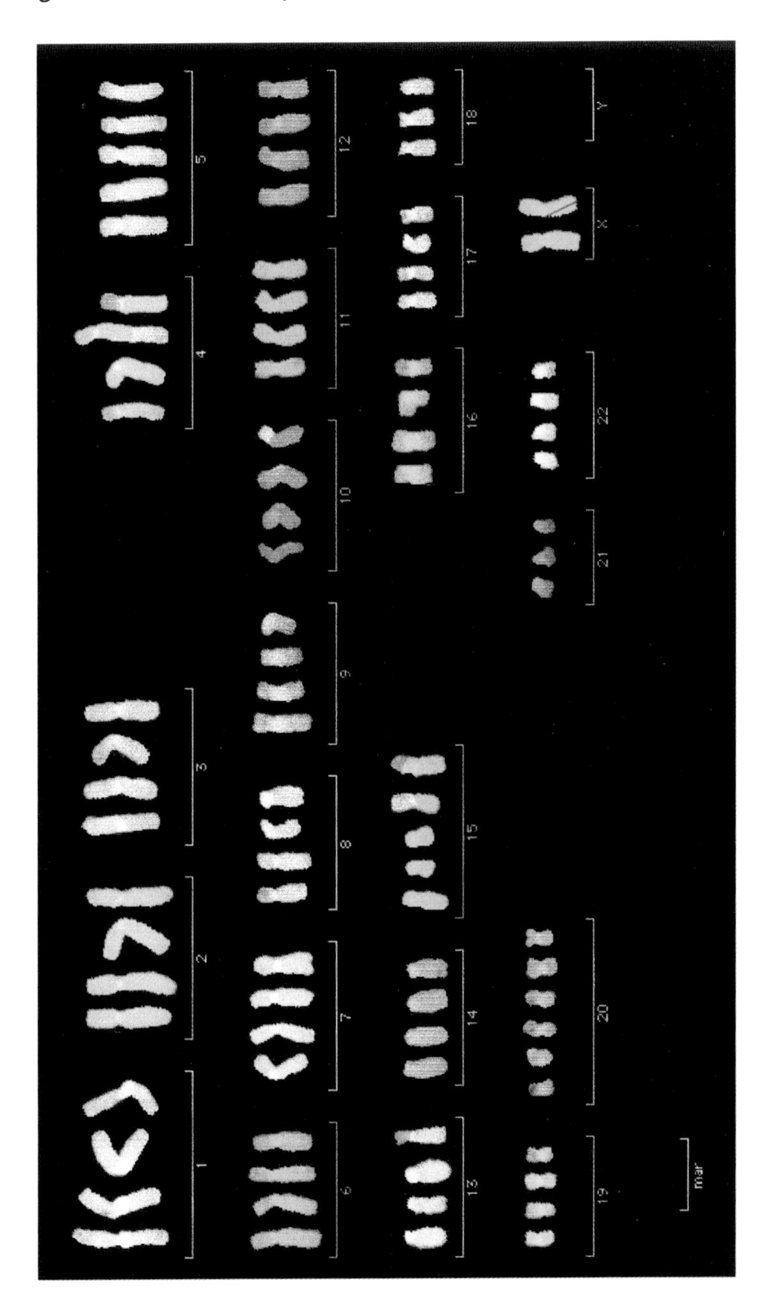

Fig. 1. The M-FISH karyotype for the bladder cancer cell line EJ28.

3. The sequence in which the fluorochromes are captured is also vital. Most default settings in MFISH software will capture the DAPI channel last, and it is important that this sequence is followed, as "bleed" through the DAPI filter can fade several of the other fluorochromes.

4. Notes

1. Owing to the complex nature of the MFISH probe and reaction, all abnormalities detected by this technique should be confirmed with conventional dual/triple-color chromosome painting. A recent study performed in our laboratory compared MFISH and SKY results on prostate carcinoma cell lines *(8)*. This study clearly suggested that these two techniques may characterize chromosome aberrations in significantly different ways, in particular when identifying complex karyotypic changes. Confirmatory painting may not be required where MFISH has detected whole arm changes or rearrangements involving relatively large segments of chromosome material, but is definitely appropriate when investigating the nature of complex chromosome markers.

Acknowledgments

I would like to thank all the members of the Cytogenetics Unit at St. Bartholomew's Hospital, London, for their experience and advice, in particular to Debra Lillington for her help with the construction of this document. I would also like to thank the Orchid Cancer Appeal and the Imperial Cancer Research Fund for their continual support of this work.

References

1. Speicher, M. R., Gwyn-Ballard, S., and Ward, D. C. (1996) Karyotyping human chromosomes by combinatorial multi-fluor FISH. *Nat. Genet.* **12,** 368–375.
2. Schrock, E., du Manoir, S., and Veldman, T. (1996) Multicolor spectral karyotyping of human chromosomes. *Science* **273,** 494–497.
3. Veldman, T., Vignon, C., Schrock, E., Rowley J. D., and Ried, T. (1997) Hidden chromosome abnormalities in hematological malig-

nancies detected by multicolour spectral karyotyping. *Nat. Genet.* **15,** 406–410.

4. Padilla-Nash, H. M., Nash, W. G., Padilla, G. M., et al. (1999) Molecular cytogenetic analysis of the bladder carcinoma cell line BK-l0 by spectral karyotyping. *Genes Chromosomes and Cancer*, **25,** 53–59.
5. Adeyink, A., Kytola, S., Mertens, F., et al. (2000) Spectral karyotyping and chromosome banding studies of primary breast carcinomas and their lymph node metastases. *Int. J. Mol. Med.* **5,** 235–240.
6. Pan, Y., Kytola, S., Farnebo, F., et al. (1999) Characterization of chromosomal abnormalities in prostate cancer cell lines by spectral karyotyping. *Cytogenet. Cell Genet.* **87,** 225–232.
7. Beheshti, B., Beheshti, B., Karaskova, J., Park, P. C., Squire, J. A., and Beatty B. G. (2000) Identification of a high frequency of chromosomal rearrangements in the centromeric regions of prostate cancer cell lines by spectral karyotyping. *Mol. Diagn.* **5,** 23–32.
8. Strefford, J. C., Lillington, D. M., Young, B. D., and Oliver, R. T. D. (2001) The use of multicolor fluorescence technologies in the characterization of prostate carcinoma cell lines: a comparison of multiplex fluorescence in situ hybridization and spectral karyotyping data. *Cancer Genet. Cytogenet.* **124,** 112–121.

17

Some Difficult Choices in Cytogenetics

John Swansbury

1. Research or Routine

It is rare for a full-fledged malignancy cytogenetics service to be started as a result of a policy decision and business plan. More often it grows from a small beginning: perhaps just one or two research assistants working on the particular interest of an oncologist, or perhaps one or two people in a preexisting general cytogenetics laboratory specializing in malignancy cytogenetics instead of the malignancy samples being shared equally between everyone. Because genetic studies of many types of malignancy are still in their early stages, then the research element is often prominent. As the findings are published and confirmed and become clinically useful, then research funding is likely to become unobtainable as the work is deemed to have become a service and should therefore be funded in the same way as other well-established clinical services such as hematology and biochemistry.

The change in work ethic is not always an easy one. It can also be difficult to combine both aspects equally in a single laboratory, when studies of one type of malignancy are still regarded as being research yet studies of another type are deemed to be routine. Research and routine are not entirely exclusive; indeed, every

From: *Methods in Molecular Biology, vol. 220: Cancer Cytogenetics: Methods and Protocols*
Edited by: John Swansbury © Humana Press Inc., Totowa, NJ

research unit should provide a result to the clinician who supplied the sample, and every routine, diagnostic service will come across unusual cases that should be investigated in greater depth with a view to publication. However, the philosophy of research and the ethos of service usually do not mix comfortably. The priority of the former is in novel discovery, often requiring prolonged, multiple studies of the same sample, and using the most up-to-date technologies; the priority of the latter is to provide the clinician with a reliable result quickly enough to be used for the patient's benefit.

This tension between research and routine is as true for the individual as for the laboratory; someone who has started his or her working life in a research environment can find it difficult to get used to the different pressures of a routine clinical service. In the author's experience, some cytogeneticists find the transition to be an uncomfortable one, and the adjustment can take a long time. There is a world of difference between the disposition needed to be innovative in experimenting and developing new ways of investigating the genetic abnormalities in cancer, and the disposition that will efficiently and systematically analyze a case to a prescribed level, produce a report, close that study, and then move on to the next case.

As a career, the world of genetics research offers scope for major discoveries that can benefit mankind; however, it can be an uncertain way of life, dependent on winning grants every few years. The relative security of employment in a service post in a routine laboratory is a job of no less value to humanity.

1.1. Research

Most research work depends on obtaining funding from grant-awarding organization, and there is usually strong competition for the funds available. To secure financial support it is necessary to be able to show proven expertise, past success in similar studies, and a well prepared plan to develop a new technique or study a malignancy that has not already been thoroughly investigated in the same way. Being the second to make a discovery is usually not good enough to win further grants, even though almost every discovery must be only tentative until it has been confirmed. Therefore, research-funded laboratories need to be at the forefront of innovation, and will use the latest and most sophisticated techniques. A successful application for grants is likely to result in the most up-to-date equipment that is beyond the budget of the average routine laboratory. If the proposal is to identify and isolate new genes relevant to the oncogenesis of a particular type of malignancy, then it is possible that rather little effort will be put into conventional cytogenetic studies, as it can take several weeks or months to obtain enough divisions from a solid tumor. Instead, it is likely that more effort will be put into comparative genomic hybridization (CGH), to identify chromosome regions of gain or loss, fluorescence *in situ* hybridization (FISH), using a series of contiguous probes to focus on selected regions, then microarrays, gene sequencing, or gene product analysis. Much time may be spent on highly detailed studies of relatively few samples.

Some research units develop a reputation for requesting particular types of samples and then never giving any feedback to the provider. It is a courtesy to provide a prompt report of the results of investigations, and to acknowledge the source of material in any publications. It is also important to remember that there is an increasing public interest in genetics and ethics; permission may need to be sought from the patient or his or her family before using clinical material for research, and it is wise to have a written agreement about what data will be reported back to the clinician and what will be passed on to the patient. The formal report should always contain a reminder that results obtained in a research environment should be used with caution; they may not have been obtained in a way that has been subject to the thorough testing and strict quality controls that apply to routine clinical diagnostic service.

In illustration of this, a widely used FISH probe for detecting a split of a gene commonly involved in acute monocytic leukemias was eventually revealed to cover only 80% of the gene; any breakpoint in the gene but outside the area covered by the probe was missed. This essential information was not so widely known, and some users were perplexed that a FISH study using this probe gave normal results for some cases that were abnormal by molecular techniques. The original researchers who produce a probe may be aware of any limitations, but the commercial suppliers who market it for routine use are unlikely to emphasize them.

1.2. Routine

If the laboratory is going to support a routine diagnostic malignancy cytogenetics service, then the focus will be on those techniques that have already been shown to identify reliably well known, clinically relevant genetic abnormalities. There is unlikely to be time to spend on developing new equipment; it is more likely that the laboratory will need to use equipment, reagents, and techniques that have already been field tested and can be trusted to provide a reliable result. There may also be little time to spend on in-depth investigations of unusual cases, when there is pressure to provide a rapid result for a large number of samples. However, time and effort spent on investigating and publishing interesting cases will serve to benefit other patients elsewhere and in the future.

2. Choice of Technologies

When DNA probes were developed for detecting the Philadelphia chromosome translocation, t(9;22)(q34;q11) ***(18)***, it was supposed that the labor-intensive cytogenetic screening for this abnormality (and similarly for other translocations) would become redundant. For a short while, the new approaches were deemed to be in competition with conventional cytogenetics. However, it is now abundantly clear that all the techniques are best used in a complementary way, selecting the most appropriate type to suit a particular clinical need. Although this is undoubtedly beneficial for the patient, it does create a dilemma for the cytogenetics laboratory: how to provide the best possible service without making it prohibitively expensive. In most centers financial constraints mean that hard decisions have to be made about how to maximize the results while minimizing the costs. This is particularly acute in centers that depend entirely on clinical work for their income and are not able to benefit from coexisting research or academic activities.

The number of technologies available to the cytogenetics laboratory is rapidly increasing. Each has its strengths and weaknesses that make it more suited to particular applications, ranging from preliminary studies of newly studied malignancies through to well established routine analyses used in a clinical setting. Whether setting up a new laboratory or seeking to expand an existing cytogenetics service, it is very likely that there will be limits on the funding and manpower available. The following comments are intended to help in decisions about the best mix of the most appropriate and cost-effective techniques to adopt. The following comments are intended to help in decisions about the best mix of techniques to adopt so as to be able to meet the requirements of those who send samples to the laboratory. The techniques are described in broad terms; as the newer ones continue to be developed and expanded, likewise the details will change.

2.1. Conventional Cytogenetics

For diagnostic work, a conventional cytogenetic study is still the most efficient and cost-effective way of examining the whole karyotype at once: all kinds of abnormalities can be detected if the affected chromosome part is large enough, including gains, losses, translocations, inversions, deletions, and duplications. However, it takes training and experience to become proficient in recognizing these, especially the more subtle abnormalities.

For follow-up studies and for diagnostic studies of tissues that are not composed entirely of malignant cells, a major drawback with conventional cytogenetics is the amount of time it takes to analyze each division, often limiting the number that can be analyzed. If 30 divisions are analyzed then it should be possible to detect a clone involving as low as 5% of divisions in 95% of cases. If the number of clonal divisions in a sample is < 5%, there is a much lower probability that it will be detected unless many more divisions are analyzed. Consequently, clones are less likely to be detected when treatment has started or in conditions where the clone is only slowly emerging.

Many of the other limitations of conventional cytogenetics studies were listed in Chapter 2 in the context of hematological malignancy. In addition, some solid tumors have cytogenetic abnormalities that are apparently identical and yet have different gene rearrangements. For example, there is a t(12;22)(q13;q12) in myxoid liposarcoma and in clear cell sarcoma. A conventional cytogenetic study would not distinguish between these two diagnoses, and it would be necessary to use FISH or reverse transcription-polymerase chain reaction (RT-PCR) to see whether it was the *CHOP* or the *ATF1* gene at 12q13 that had been fused to the *EWS* gene on 22q12. Similarly, a cytogenetic study that detected a t(X;18)(p11;q11) in a synovial sarcoma would not show whether it was the *SSX1*, *SSX2*, or *SSX4* gene at Xp11 that was involved, which may be important because the *SSX1* seems to be associated with a worse prognosis ***(1)***.

The major equipment costs include a good, comfortable microscope, with a 10× objective lens for screening and a 100× oil-immersion objective lens for analysis. Ideally, each cytogeneticist should have a microscope for their own use, which they can adjust to suit his or her comfort. If the laboratory can afford to get one, an automated karyotyping system has several advantages in terms of helping to reveal some of the subtle abnormalities, facilitating checking, simplifying training, and creating electronic images for keeping records. If an automated system is not available, then the microscope should have a camera attached so that photographs of suitable metaphases can be taken and printed; the chromosomes can then be cut out and arranged in pairs to produce a karyogram—the formal arrangement of chromosomes as shown in many of the illustrations in this book. With experience it becomes possible to analyze most metaphases directly down the microscope, either by counting the number of chromosomes present and then systematically scoring for the presence of each, or else by drawing a sketch of the location of each chromosome. However, if a complex clone is found, a full analysis usually requires the preparation of a formal karyogram.

The reagents used for culturing and processing samples form a small part of the overall cost of performing a cytogenetic study, in contrast to those in the other technologies. The major part of the expense is attributable to staff costs, as so little of the work is amenable to automation.

Not everyone is suited to karyotype analysis. In the author's experience, there have been many trainees who have been well qualified in other respects, and holding senior positions in other types of medical work, but who have been unable to achieve a secure grasp of the normal chromosome banding patterns. After 6 mo of effort, they were still unable to produce consistently accurate karyotypes. Because other trainees can reach a good measure of proficiency within 1 or 2 mo, there would appear to be an inherent difference in pattern recognition ability. This variation in aptitude needs to be recognized before making the assumption that anyone can be trained to become a competent cytogeneticist. Other types of analysis such as FISH can be easier to learn, but even then an ability to be able to identify all the chromosomes by their banding pattern is often needed to interpret the FISH results.

2.2. FISH

As described in Chapters 13–16, there is a great (and growing) variety of DNA probes that are available for FISH analysis. These include whole chromosome paints, part-chromosome paints, alpha satellite probes for identifying centromeres, telomeric probes, gene-specific probes, and probes for parts of genes. With more sophisticated microscopes, cameras, and multiplex FISH (MFISH) or spectral karyotyping (SKY) computer systems, it is also possible to identify all chromosomes simultaneously.

Many of the FISH probes still require the availability of good-quality metaphase spreads, just as for conventional cytogenetics. However, some, in particular the alpha satellite and gene-specific probes, can be used in interphase; this makes them particularly valuable for overcoming the requirement for divisions that is one of the major limitations of conventional cytogenetics. Such analyses have resolved some of the mysteries that had beset cytogenetics studies. For example, in an instance when all the dividing cells in a patient who had relapsed after bone marrow transplant were still donor, a FISH study showed that the majority of the nondividing cells were of recipient origin; it was not necessary to suppose that the relapse had occurred in donor cells.

Compared to a conventional cytogenetic study, the cost of reagents is much higher. At the time of writing, the cost of just the probe needed for one slide for an MFISH study can be as much as the entire cost of a conventional cytogenetic study. The prices of probes will probably reduce in time, but are likely to remain a significant factor in the implementation of a FISH service. The probes also deteriorate after some months, and the hybridization efficiency can be affected if the processing conditions are not perfectly correct.

It is generally much quicker to check interphase or metaphase cells for a FISH result than to make a complete cytogenetic study. Therefore, 100–200 interphase cells or more can be scored if necessary, giving FISH a hypothetical clone detection rate of approx 0.5%. It has been shown that this intermediate level of sensitivity can be more useful, clinically, than the greater sensitivity of some molecular assays; for example, persisting low levels of *PML-RARA* fusion detected by RT-PCR have been found in some patients with acute myeloid leukemia (AML) M3 in long-term remission ***(2,3)***, but the detection of any fused *PML-RARA* signals by FISH is associated with imminent relapse ***(4)***.

However, in practice, a >0.5% threshold has to be set for interphase FISH, as there can be technical reasons for spurious results. For example, if screening for the presence of a clone with monosomy 7 using just a chromosome 7 centromeric probe, there are several reasons why a nucleus may have only one signal, including failure of one chromosome 7 to hybridize, or the superimposition of two signals. These events can produce 2% or more cells with only one signal, so this would mask the presence of a low-level monosomy 7 clone. In such circumstances, some of the uncertainty can be overcome by simultaneously including other probes, such as one also located on the 7s but well away from the centromere.

As well as its value in testing interphase cells, FISH is frequently used to resolve cryptic or complex chromosome abnormalities in metaphase cells, and to identify the exceptional cases where genes have been rearranged in the absence of any detectable exchange of chromosome material (*see* Chapter 3, **Subheading 4.3.**). In addition, it is being increasingly recognized that a substantial proportion of what appear to be balanced translocations are associated with submicroscopic deletions that are clinically significant; these are rarely suspected by conventional cytogenetics but are detected by modern FISH probes *(5)*.

The ability of MFISH or SKY to identify all the chromosomes in a metaphase spread is attractive, and their use has led to the resolution of complex chromosome rearrangements and the discovery of unexpected abnormalities. However, the reagents for these technologies are expensive, and the analysis is very time-consuming. In practice, the results of an MFISH or SKY analysis are often best regarded as being preliminary, and they should be confirmed by using dual- or triple-color FISH. Also, most chromosome inversions and small deletions will tend to go unrecognized. Finally, the results obtained by MFISH are not always identical to those obtained by SKY *(6)*.

CGH is directed toward the detection of gains or losses; it does not detect balanced translocations. It is particularly good for detecting gene amplification: Tumor cells may contain large numbers of double minutes or long homogeneously staining regions (HSRs) which cannot usually be identified from their morphology. CGH has the advantage of not needing divisions, but it does require that the sample has a high proportion of clonal cells. Interpretation of the results is not always clear, and false positives occur *(7)*.

In general, dual- and triple-color FISH is mostly used to detect the presence of specific abnormalities that are already known or suspected to be present, and so its use at diagnosis tends to be limited to screening for a limited number of specific abnormalities. It is less likely than conventional cytogenetics to discover translocation partners, variant translocations, and co-occurring secondary abnormalities.

This fact tends to be overlooked by those who advocate using FISH alone to screen all new cases, yet the missing information can be of clinical importance. For example, using a *MLL* DNA probe will identify those cases of leukemia in which the *MLL* gene is split, but may not reveal which of the >50 known translocation partners is involved. However, it is the translocation partner that is most useful in defining the diagnosis and prognosis for these patients. Similarly, FISH using a probe for the EWS gene would identify more cases than would be detected by a conventional cytogenetic study, but other studies, either conventional cytogenetics or FISH studies using other probes, would be needed to show which of the at least nine translocation partners was involved.

2.3. Molecular Analyses

As explained in the introduction, this book does not include any description of the techniques for molecular analyses such as PCR or RT-PCR. However, some comparisons between them and conventional cytogenetics or FISH are worth mentioning here.

PCR and RT-PCR have two particularly powerful attributes: They do not need dividing cells, and they are far more sensitive—the hypothetical clone detection rate can be below 0.0001%. Other molecular assays can be used on gene products, rather than gene structure, which may be more relevant to the in vivo situation.

The greater sensitivity of PCR-type assays introduces its own considerations. For example, the slightest amount of cross-contamination can give a positive result. Also, the test may actually be too sensitive for clinical needs: A patient in remission may continue to give positive results for a particular gene rearrangement without it necessarily meaning that relapse is likely. As mentioned previously and in Chapter 3, there are patients who have been in remission after treatment for AML for several years in whom molecular evidence of t(8;21)(q22;q22) or t(15;17)(q24;q21) can still be found without it indicating an adverse clinical consequence.

Quantitative PCR assays have been devised and are still being developed. These may not give particularly accurate values for the level of clone present, but are an advance on a simple positive/negative result. Data are accumulating about the levels that become clinically relevant.

Another factor to be considered is that expression of some of the typical leukemia- and lymphoma-associated gene rearrangements has been found at low levels in some people who had no evidence of disease ***(8–12)***. In some cases it can be shown that the fusion transcripts obtained from normal people are different from those obtained from patients, if the right test is used ***(11)***.

As well as being more sensitive, molecular methods are generally highly specific: If the breakpoints in a gene are outside the are covered by the primers, or if a different partner gene is involved, then the rearrangement may go undetected and a false-negative result obtained.

Lastly, despite such great sensitivity a negative result still does not prove that all clonal cells are absent. The author's laboratory studied a child with Ph+ ALL who had a good response to treatment; the clone became undetectable by conventional cytogenetics, and a series of quantitative RT-PCR assays tracked levels down to fewer than 10 transcripts in 690,000. However, within 2 wk of this result, the disease had relapsed, and the patient died shortly afterwards.

As with most clinical investigations, it is unwise to depend entirely on the results of one test. Wherever possible, a positive result obtained by a molecular genetic study should be confirmed with another test, and a negative result interpreted with caution ***(13,14)***.

3. Summary

In making a selection of features of these technologies, it is inevitable that some will be omitted that other cytogeneticists feel should have been included. The author could probably justifiably be accused of bias. However, based on experience in a laboratory that has used almost every type of assay mentioned in this chapter, the following opinions are offered about their current value in providing a routine malignancy cytogenetics service:

1. The foundation is still a conventional cytogenetic study, preferably with the use of an automated karyotyping system.
2. Added to this, there should be the capability of performing FISH studies using chromosome paints and gene-specific probes. Cytogenetics and FISH form a powerful partnership when backed by experienced cytogeneticists. MFISH or SKY are also useful if the laboratory can afford the considerable extra expense. CGH and fibre FISH are generally better suited to research projects, and at present have few applications in a routine diagnostics service.
3. At present, molecular methods such as RT-PCR mostly tend to produce results that have a greater need of confirmation by other techniques before they can be used for clinical management.

References

1. Panagopoulos, I., Mertens, F., Isaksson, M., et al. (2001) Clinical impact of molecular and cytogenetic findings in synovial sarcoma. *Genes Chromosomes Cancer* **31,** 362–372.
2. Tobal, K., Saunders, M. J., Grey, M. R., and Yin, J. A. (1995) Persistence of RAR alpha-PML fusion mRNA detected by reverse transcriptase polymerase chain reaction in patients in long-term remission of acute promyelocytic leukemia. *Br. J. Haematol.* **90,** 615–618.
3. Diverio, D., Rossi, V., Avvisati, G., et al. (1998) Early detection of relapse by prospective reverse transcriptase-polymerase chain reaction analysis of the *PML/RAR*alpha fusion gene in patients with acute promyelocytic leukemia enrolled in the GIMEMA-AIEOP multicenter "AIDA" trial. GIMEMA-AIEOP Multicenter "AIDA" Trial. *Blood* **92,** 784–789.
4. Zhao, L., Chang, K. S., Estey, E. H., Hayes, K., Deisseroth, A. B., and Liang, J. C. (1995) Detection of residual leukemic cells in patients with acute promyelocytic leukemia by the fluorescence in situ hybridization method: potential for predicting relapse. *Blood* **85,** 495–499.
5. Kolomietz, E., Al-Maghrabi, J., Brennan, S., et al.(2001) Primary chromosomal rearrangements of leukemia are frequently accompanied by extensive submicroscopic deletions and may lead to altered prognosis. *Blood* **97,** 3581–3588.
6. Strefford, J. C., Lillington, D. M., Young, B. D., and Oliver, R. T. D. (2001) The use of multicolor fluorescence technologies in the characterization of prostate carcinoma cell lines: a comparison of multi-

plex fluorescence in situ hybridization and spectral karyotyping data. *Cancer Genet. Cytogenet*, **124,** 112–121.
7. Barth, T. F., Benner, A., Bentz, M., Dohner, H., Moller, P., and Lichter, P. (2000) Risk of false positive results in comparative genomic hybridization. *Genes Chromosomes Cancer* **28,** 353–357.
8. Biernaux, C., Loos, M., Sels, A., Huez, G., and Stryckmans, P. (1995) Detection of major *bcr-abl* gene expression at a very low level in blood cells of some healthy individuals. *Blood* **86,** 3118–3122.
9. Muller, J. R., Janz, S., Goedert, J. J., Potter, M., and Rabkin, C. S. (1995) Persistence of immunoglobulin heavy chain/c-*myc* recombination-positive lymphocyte clones in the blood of human immunodeficiency virus-infected homosexual men. *Proc. Natl. Acad. Sci. USA* **92,** 6577–6581.
10. Dolken, G., Illerhaus, G., Hirt, C., and Mertelsmann, R. (1996) *BCL-2/JH* rearrangements in circulating B cells of healthy blood donors and patients with nonmalignant diseases. *J. Clin. Oncol.* **14,** 1333–1244.
11. Marcucci, G., Strout, M. P., Bloomfield, C. D., and Caliguri, M. A. (1998) Detection of unique *ALL1 (MLL)* fusion transcripts in normal human bone marrow and blood: distinct origin of normal versus leukemic *ALL1* fusion transcripts. *Cancer Res.* **58,** 790–793.
12. Uckun, F. M., Herman-Hatten, K., Crotty, M. L., et al. (1998) Clinical significance of *MLL-AF4* fusion transcript expression in the absence of a cytogenetically detectable t(4;11)(q21;q23) chromosomal translocation. *Blood* **92,** 810–821.
13. Hunger, S. P., and Cleary, M. L. (1998) Commentary: what significance should we attribute to the detection of *MLL* fusion transcripts? *Blood* **92,** 709–711.
14. Tajiri, T., Shono, K., Fujii, Y., et al. (1999) Highly sensitive analysis for N-myc amplification in neuroblastoma based on fluorescence in situ hybridization. *J. Pediatr. Surg.* **34,** 1615–1619.

18

Introduction to the Analysis of the Human G-Banded Karyotype

1. The ISCN

There is an internationally agreed system for describing the banding pattern of chromosomes, such that if an abnormality is accurately described in one laboratory then it can be recognized in another. This is known as the ISCN, the International System for Human Cytogenetic Nomenclature. Since its first appearance in 1965, it has been tried, tested, and modified, and the 1995 edition remains the standard version in current use *(1)*. This is an essential reference for the definition of cytogenetic abnormalities. Within the ISCN are the formal descriptions of how to describe chromosome bands and abnormalities. There is also a schematic representation of the human karyotype, and several illustrations of karyotypes of normal chromosomes, stained in different ways. Every cytogeneticist needs to become familiar with the correct way of describing chromosomes and their abnormalities.

The 1995 edition of the ISCN included the first system for describing the results of fluorescence *in situ* hybridization (FISH) studies. In this author's opinion, the FISH nomenclature system generally works well enough for simple abnormalities, but is less successful for complex ones. After a few more years of experience

From: *Methods in Molecular Biology, vol. 220: Cancer Cytogenetics: Methods and Protocols*
Edited by: John Swansbury © Humana Press Inc., Totowa, NJ

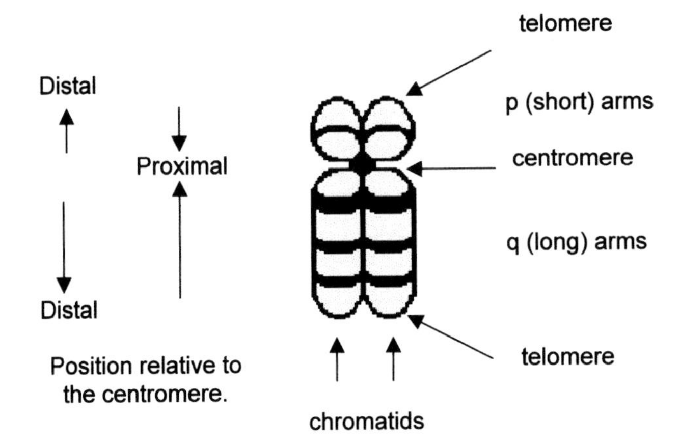

Fig. 1. Schematic diagram of a chromosome, to illustrate some of the descriptive terms.

with FISH in malignancy have been obtained, it is hoped that some improvements will be introduced.

2. The Basics of Chromosome Structure

As shown in **Fig. 1**, metaphase chromosomes are composed of paired chromatids that are joined at a centromere. During cell division, the centromere attaches to the spindle, a temporary fibrous structure that pulls the chromatids apart so that each goes to a new daughter cell. The location of the centromere is used in a description of the chromosome appearance: A metacentric chromosome has the centromere near the middle of the chromatids, and an acrocentric chromosome has the centromere near the end; submetacentric is anywhere in between.

The chromatids are composed of a long, coiled molecule of DNA, surrounded by proteins. In interphase (between cell division), the DNA is in an extended form, contained within the cell nucleus, and no chromosome structure can be seen by light microscopy. Before the cell can divide, the DNA must be duplicated. Then, during the first stage of cell division, *prometaphase*, the DNA is coiled and recoiled into a condensed form, with each chromosome becoming a discrete, microscopically visible unit. At *metaphase*, the chromosomes are shortest and are most easily analyzed; the nuclear membrane breaks down, so that each chromosome is released into the cell cytoplasm. During the next stage, *anaphase*, the centromeres attach to the spindle and divide, and each chromatid is drawn along the spindle fibers to opposite ends of the cell. During the last stage, *telophase*, the DNA unwinds, two new nuclear membranes reform, and the cytoplasm divides, forming two new daughter cells. For cytogenetic studies, an agent, usually colcemid or colchicine, is used to destroy the spindle, so that the chromatids do not separate. If there is not quite enough colcemid, then the spindle starts to form and the centromeres attach to it but do not separate; the resulting ring of chromosomes can be recognized and indicates that more colcemid should be used to harvest subsequent cultures. Sometimes the short arms of some of the chromosomes 13, 14, 15, 21, and 22 are linked, or form a small ring. This is not due to lack of colcemid but arises because these chromosomes are associated with the formation of the nucleolus during interphase.

3. Chromosome Nomenclature

Chromosomes are designated by a number corresponding to decreasing size, with the exception that chromosome 21 is smaller than 22. There is a historical reason for this, linked to the description of a constitutional extra no. 21 chromosome in Down syndrome before the discovery of banding that made all chromosomes identifiable.

Size alone is not sufficient to identify all the chromosomes, as many are of similar size and shape. Full analysis depends on knowing the banding pattern. Although bands can be produced in different ways, the banding pattern is consistent.

Each chromosome arm is first divided into regions using any conspicuous dark or light bands; each region is then subdivided, as on longer chromosomes the bands can often be resolved into smaller units. With extended chromosomes, further precision is possible. As formally described in the ISCN, numbers have been assigned to all the chromosome bands. Thus the breakpoint on chromosome 9 involved in the Philadelphia translocation is described as being at 9q34.1 ("nine Q three four point one").

In most countries, the preferred method of banding chromosomes produces what is called a G-banded or GTG-banded pattern, so named because it was originally produced by using trypsin and Giemsa stain. In some European countries, the preferred method of banding uses fluorescent stains such as quinacrine or acridine orange, which do not require pretreatment with trypsin. These stains produce what is known as R-banding, and this is generally the reverse of the G-banding pattern, such that a dark band in one is in the same position as a light band in the other. Although it is technically easier to produce R-bands, many people find them more difficult to analyze, largely because the reflected fluorescent light is faint, diffuse, difficult to photograph, and tends to fade rapidly.

In this chapter, only the G-banding pattern is described.

4. Learning How to Analyze

Before any abnormality can be recognized, the cytogeneticist must be familiar with the normal appearance of human chromosomes. Memorizing and learning to recognize the banding patterns of each chromosome is easier for some people than for others, but it usually takes at least 2 mo of regular, consistent practice to become proficient. It is usual to start with photographs of good quality, normal metaphases; the chromosomes are cut out, arranged into pairs, and stuck onto a card (*see* **Fig. 2** in Chapter 12). The example karyotypes in the ISCN can be used as a reference to determine the identity of each pair, or the trainee may follow a previously prepared and corrected karyotype. The chromosomes are normally placed with the short arms uppermost. If an abnormal chromosome is found, it is usual to place it to the right of the pair. If a chromosome contains a pericentric inversion, then the chromosome should be aligned so that the telomeres are in the correct orientation. As the trainee completes each karyotype it should be checked promptly to ensure that no consistent errors are being made.

Some chromosomes are easier to recognize than others, and it can be helpful to start with these and master them before attempting the more difficult ones. A guide to the features of each chromosome that help in identifying it is given in **Subheading 5.** Particular reference is made to those features that may be used to distinguish between similar chromosomes.

Learning the human karyotype in malignancy has an extra challenge: With constitutional studies, chromosomes with good morphology can be consistently obtained in almost all cases; in malignancy, however, the chromosome morphology can vary widely, from very short, with almost no bands at all, up to elongated, prometaphase chromosomes with nearly 1000 bands per haploid set. The malignancy cytogeneticist therefore has to become familiar with the appearance of chromosomes at all levels. The guide in **Subheading 5.** uses both short and long chromosomes in illustrations; however, shorter chromosomes with about 300 bands per haploid set are as much as can be achieved in many malignancy studies.

There is a class of abnormality that is probably more frequently associated with poor quality chromosomes than any other, and that is the "high hyperdiploid" class that is found in acute lymphoblastic leukemia. These clones typically have approx 55 chromosomes, and in many cases they are short and poorly spread, with hardly any banding, despite every technical effort made to improve their appearance. Sometimes it is impossible to analyze them fully, and the clone may have to be defined simply by counting the chromosomes and demonstrating a consistent modal number. An inexperienced cytogeneticist may miss the clone in these circumstances, by disregarding the unanalyzable divisions. When starting to study a new patient it is perfectly acceptable to analyze a few good-morphology divisions in case these turn out to be clonal; however, it is essential to attempt to analyze (or even partly analyze) any poor-morphology divisions that may be present.

Once a trainee has been found to be consistently accurate in analyzing metaphase chromosomes by cutting them out and physically pairing them, then the ability to analyze down the microscope should be developed. Even in laboratories that have an automated karyotyping system, much of the routine analysis of follow-up samples is done in this way. It is useful to have paper and pencil beside the microscope so that a sketch of the location of the chromosomes can be made.

The location of each metaphase should be recorded so that it can be found again if necessary. This is particularly important if an abnormality is subsequently recognized and the earlier divisions need to be checked to see if it was missed. Most microscopes have a vernier scale along each axis which gives an accurate indication of location. However, vernier readings do not usually correspond between microscopes. If a metaphase is to be found on another microscope, then a different system has to be used: one way is with an industry-standard, marked slide, known as an England Finder. For this to work, it is necessary to know which way the slide was placed on the microscope stage (label to left or right) and which sides of the slide holder formed the fixed edge.

5. A Guide to Distinguishing Between Normal Human Chromosomes

In malignancy cytogenetics the chromosome morphology can be very poor, and sometimes the chromosomes obtained bear little resemblance to the diagrammatic illustrations in the ISCN. Some chromosomes have a characteristic banding pattern that is usually easily identified, but others can be very difficult to distinguish. Chromosomes 4 and 5, for example, and chromosomes 14 and 15 can very closely resemble each other. The description in **Figs. 2** to **18**, therefore, is intended to give some guidance about what landmarks to use when learning how to recognize the chromosomes in the human karyotype.

6. Abnormal Chromosomes

Some chromosomal variation in size and shape is unavoidable, and is a result of the technical processing. Also, some variations in chromosome shape are inherited; this is particularly obvious in the

Chromosome 1

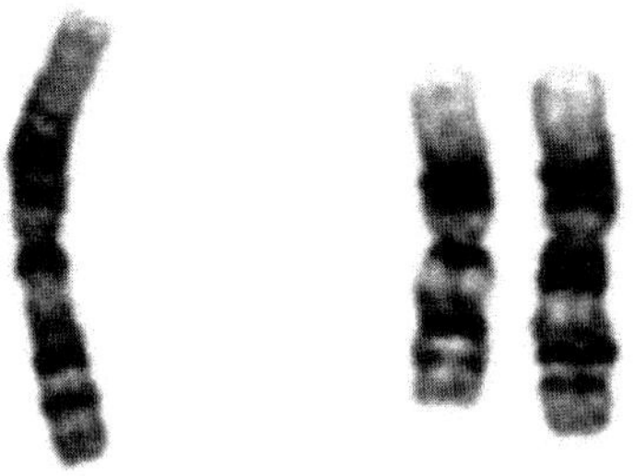

Fig. 2. Chromosome 1 is usually easy to identify but sometimes with poor morphology it is not clear which way up it should be. The dark heterochromatic region is just below the centromere. The figure also shows two chromosomes from the same patient, illustrating how much the heterochromatin can vary in size. The size is consistent in every cell for each person, as it is inherited. Chromosomes 9 and 16 also have heterochromatic regions than can vary in size, as shown in Figs. 9 and 13.

Fig. 3. Chromosome 2 has few obvious bands; it often looks uniformly dark unless the morphology is good. Here and in most of the figures, two chromosomes have been taken from different cells to show how longer chromosomes have a clearer banding pattern.

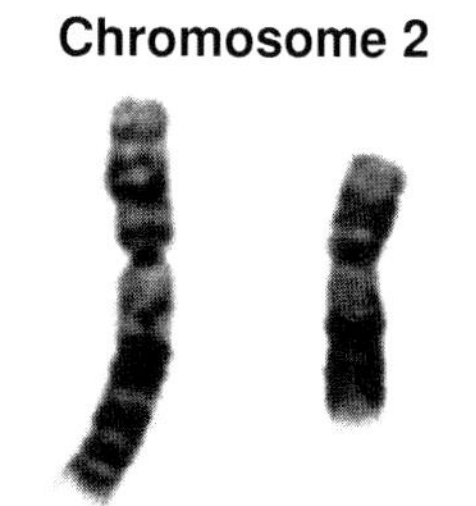

Chromosome 3

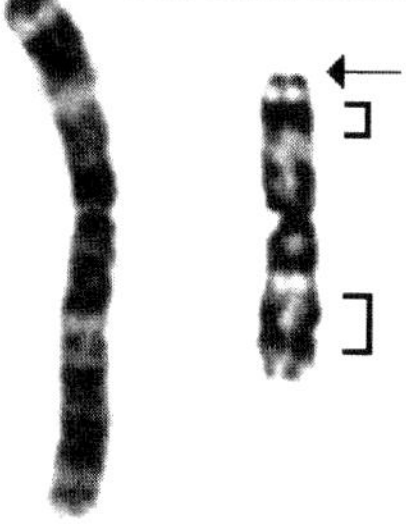

Fig. 4. Chromosome 3 is almost metacentric and it can be difficult to know which are the short arms and which are the long arms: a very small dark band is more obvious at the tip of the short arm, and the dark band toward the end of the long arm is larger than that on the short arm.

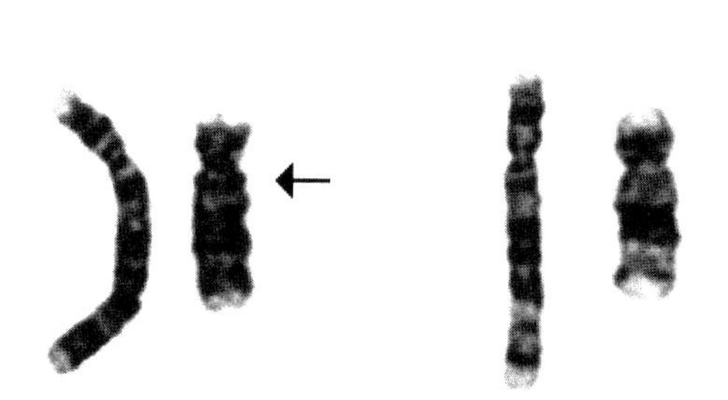

Fig. 5. Chromosomes 4 and 5 often look very similar; the 4 tends to have darker "shoulder" bands just below the centromere, and the 5 has paler bands in the long arms.

Chromosome 6

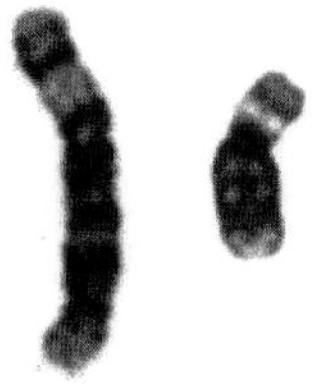

Fig. 6. Chromosome 6 has a distinctive pale band in the short arms; It can sometimes resemble a chromosome 3, but 6q does not have the distinctive pale band that is seen at 3q21-3.

Chromosome 7

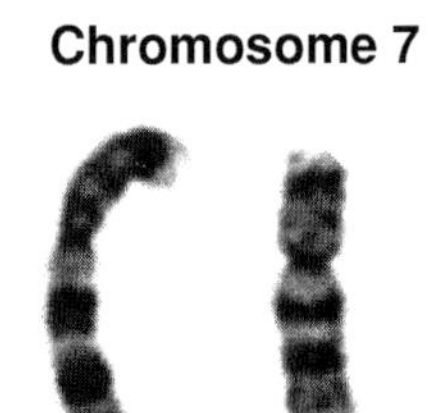

Fig. 7. Chromosomes 7 and 9 can look similar but the 7 has larger, more triangular p arms and has two dark bands on the q arms and then a third half-dark band, whereas the 9 has two dark bands and then a pale band. In short chromosomes, the 9 may appear to have no pale band in the p arms, whereas it is always evident in the p arms of a 7.

Chromosome 8

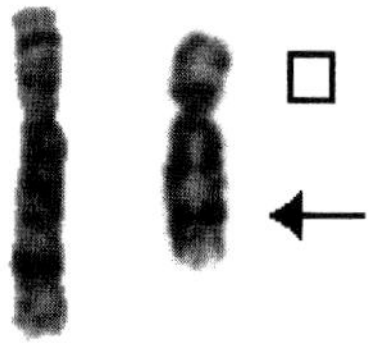

Fig. 8. Chromosome 8 can resemble 9 and 10 but has two distinguishing features: (1) there is a pair of small dark bands on each of the p arms which usually gives them a square appearance; (2) the two major dark bands on the q arms are of different intensity, the distal one being noticeably darker.

Chromosome 9

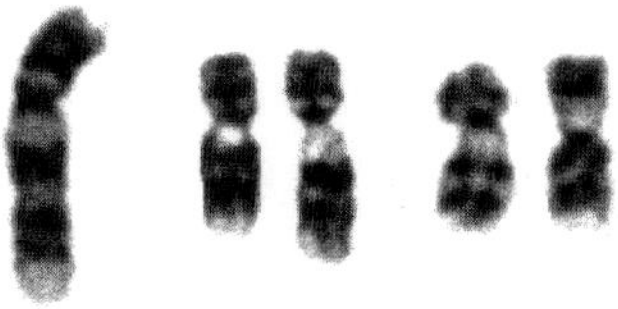

Fig. 9. Chromosome 9 has a pale heterochromatic band just below the centromere, the length of which is inherited. It can be almost invisible, and it can be longer than shown here. In 5–10% of the population, there is an inherited inversion such that this pale heterochromatic band is above the centromere, as shown on the *far right*. This is not known to have any clinical effect on the person carrying it.

Chromosome 10

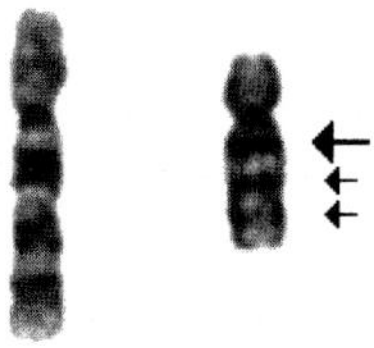

Chromosome 10 has three dark bands on the q arms, the proximal being darkest. These can resemble those of a 7 but the short arms of the 10 are much smaller.

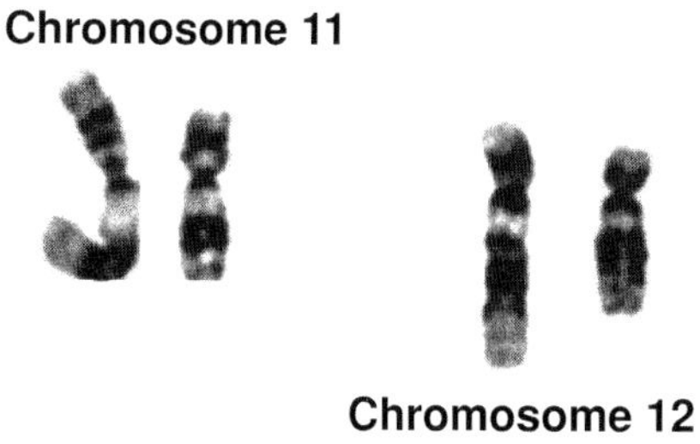

Chromosome 11

Chromosome 12

Fig. 11. Chromosomes 11 and 12 are very similar. However, the 11 looks shorter and fatter; the 12 looks long and slim and the centromere is less metacentric, resulting in shorter p arms and longer q arms. Chromosome 9 can also look similar to 11, but it has darker, more triangular short arms, and two dark bands on the long arms rather than one large one.

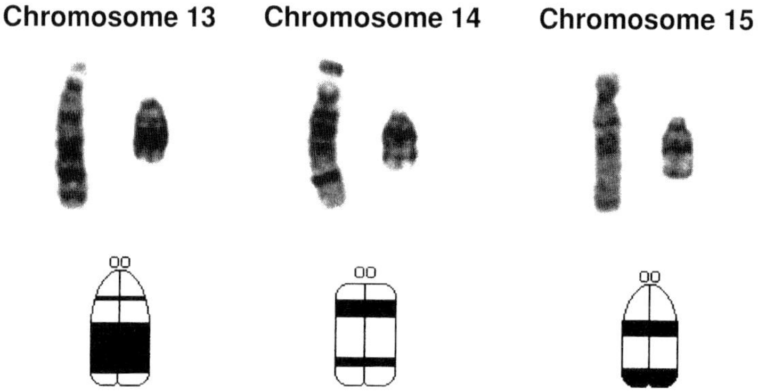

Chromosome 13 Chromosome 14 Chromosome 15

Fig. 12. Chromosomes 13, 14, and 15 all have very short p arms, and it is the q arms that are used to distinguish between these chromosomes. There are dark bands on most of the distal part of 13, sometimes fused into one large dark band. The 14 has a pair of dark bands, one near the centromere and one near the telomere, giving it a more square or rectangular shape. The 15 has a smaller, dark band further from the centromere than a 14, and most of the distal part is pale. At the end of the short arms of these chromosomes (and of chromosomes 21 and 22) there are often visible satellites, attached by almost invisible stalks. These satellites can be very small and pale, as on the 13 above; sometimes they are large and dark, as on the 14; and sometimes they are not obvious, as in the 15. They can also be duplicated. These variations are inherited, rather like the heterochromatic regions of chromosomes 1, 9, and 16, but are not consistently expressed in all divisions.

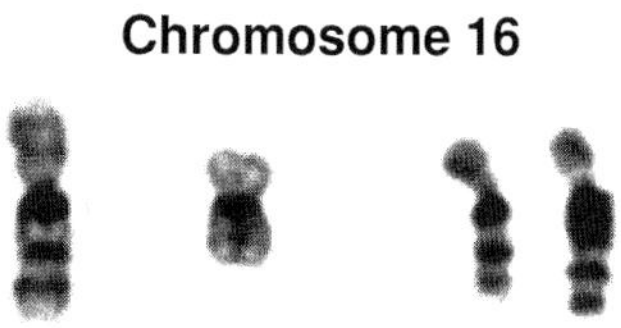

Chromosome 16

Fig. 13. Chromosome 16 has a conspicuous dark heterochromatic band just below the centromere; the size of this band is inherited, and there is a wide variation: In some chromosomes it is a tiny dot, and in others it can be as large as the rest of the q arms, as shown.

Chromosome 17

Chromosome 18

Fig. 14. Chromosomes 17 and 18 are similar but the 18 has two similarly dark bands on the long arms, whereas the 17 has a paler proximal band and darker distal band.

Fig. 15. Chromosomes 19 and 20 are similar in size and shape, but the 19 has much paler arms and a darker, larger centromere, while the 20 has small but distinct dark band on each arm, and a smaller dark centromere.

Chromosome 19

Chromosome 20

Chromosome 21

Chromosome 22

Fig. 16. Chromosome 21 has a larger dark q band and a shorter pale band; the dark band appears to be fused with the centromere. The 22 has a small dark centromere and longer, pale q arms with only a narrow dark band being evident. Satellites of various shapes can sometimes be seen at the end of the very short p arms of the 21s and 22s, as on the 13s–15s. If they are prominent, they can make a 22 closely resemble a 19.

Fig. 17. The X chromosome can often look like a 7 or 9 but has a characteristic dark band in the middle of pale short arms, one very dark band on the long arms, and no distinct pale band at the end of the long arms.

X Chromosome

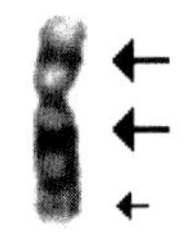

Y Chromosome

(A Yand an 18 from the same metaphase.)

Fig. 18. The long arms of the Y chromosome can vary in size from person to person, with no phenotypic effect. They can be smaller than a 21 or as large as an 18, as shown here. They are often uniformly dark and tend to lie more closely parallel than those of other chromosomes.

heterochromatic bands below the centromeres of chromosomes 1, 9, and 16. Experience of analyzing normal chromosomes will help to distinguish between this intrinsic variability and the sometimes subtle, acquired, clonal abnormalities that occur in malignancy. This subject is also mentioned in Chapter 12. Each trainee should therefore take every opportunity to study chromosome abnormalities in previously reported cases and in cases presented at meetings, so as to become familiar with their appearance. Some abnormalities, such as inv(16)(p13q22), are far easier to recognize once they have been already committed to memory.

Reference

1. ISCN (1995) *An International System for Human Cytogenetic Nomenclature.* (Mitelman, F., ed.), S Karger, Basel, 1995.

Index

Lymphoma, classifications of, 97

U

V

W

X

Y

About the Authors

Conventional Cytogenetic Techniques in the Diagnosis of Childhood Acute Lymphoblastic Leukemia
and
FISH, CGH, and SKY in the Diagnosis of Childhood Acute Lymphoblastic Leukemia

Susana C. Raimondi and Susan Mathew
St. Jude Children's Research Hospital, and University of Tennessee College of Medicine, Memphis, TN, USA.

The St. Jude Children's Research Hospital is the largest childhood cancer research center in the US, and it is recognized around the world for its treatment of children with cancer. The cytogenetics laboratory gained a worldwide reputation for its success rates and clone detection rates under its previous director, Dorothy Williams, and the high standard has been enhanced by **Susana C. Raimondi**, who has been its director since 1989. She became a Founding Fellow of the American College of Medical Genetics in 1993. Her coauthor for this book is **Susan Mathew**, who prior to joining the Department at St. Jude's had gained a wide experience of human genetics in India, Germany and New York.

Human Solid Tumors: Cytogenetic Techniques

Pietro Polito, Paola Dal Cin, Maria Debiec-Rychter and Anne Hagemeijer
The Centre for Human Genetics, University of Leuven, Leuven, Belgium.

Leuven is renowned as a major European centre for cytogenetics and molecular cytogenetics of malignancies. The Centre for Human Genetics at the University of Leuven grew from the pioneering work of Herman Van den Berghe, and is presently directed by **Anne Hagemeijer**. Under the supervision previously of **Paola Dal Cin** and more recently of **Maria Debiec-Rychter**, approximately 1000 solid tumors/year are karyotyped and submitted to FISH analysis when required. Current research projects focus on gastrointestinal stromal tumors, germ cell tumors, eye melanoma, brain tumors, some renal tumors and all mesenchymal tumors.

Chromosome Preparations from Bone Marrow in Acute Lymphoblastic Leukemia: Cytogenetic Techniques

Ann Watmore
North Trent Cytogenetics Service, Sheffield Children's Hospital Trust, Sheffield, UK

As Head of Cytogenetic Services in Hematological Malignancy, **Ann Watmore** is responsible for the running and development of diagnostic cytogenetics for both adults and children in the North Trent Region (population 1.8 million) of England. Based at Sheffield Children's hospital, Ann has worked continuously in this field since 1976, at first in research and technical development and later taking responsibility for the diagnostic service in 1991. The laboratory works in close liaison with regional clinical hematologists and clinical trials coordinators. The centre has been a driving force in the development of diagnostic cytogenetics for hematological malignancy throughout the UK via its many training courses. Ann is the current secretary of the UK Cancer Cytogenetics Group, through which all malignancy cytogenetics laboratories in the UK participate in collaborative studies.

Cytogenetic Studies Using FISH: Background

and

FISH Techniques

Toon Min
Academic Department of Haematology & Cytogenetics, The Royal Marsden NHS Trust, Sutton, Surrey, UK.

The Royal Marsden NHS Trust was founded in 1851 by William Marsden, following the death of his first wife from cancer, and it is concerned solely with the treatment and care of patients with cancer. From it grew the Institute of Cancer Research, and the two organizations retain close links, forming the largest cancer research institution in Europe. **Toon Min** previously worked in Bristol and Galway, and is responsible for the diagnostic FISH analyses done in the department.

Solving Problems in Multiplex FISH

Jon C. Strefford
CRUK Medical Oncology Unit, John Vane Science Centre, Queen Mary and Westfield College, London, UK.

Jon C. Strefford has gained cytogenetics experience by working in Bristol, Swansea, and Manchester. In 2002 Jon received a lectureship in Medical Oncology from Queen Mary and Westfield College, University of London. He currently works to identify novel cytogenetic and molecular markers in male urological tumors, in particular prostate and renal cancer. This research is funded by the Orchid Cancer Appeal, the first UK-based charity focused on improving educational awareness and research into testicular and prostate cancer.

About the Editor

John Swansbury began his career as a cytogeneticist at the Royal Marsden Hospital, Sutton, UK. After some years as a research assistant to Dr. Lorna Secker-Walker, he took up a post in the routine laboratory under the direction of the late Professor Sylvia Lawler. He has subsequently been in charge of the cytogenetics service to the hospital for over fifteen years, currently with Professor Daniel Catovsky as Head of the Department of Academic Hematology. He was a participant in the Fourth and Sixth International Workshops on Chromosomes in Leukemia, and has been a member of various committees concerned with hematology cytogenetics in the UK. He is an honorary team leader in the Institute of Cancer Research and a Fellow of the Royal College of Pathologists, London.